AF763448

Woodhead Publishing Series in Food Science, Technology and Nutrition: Number 198

Freeze-drying of pharmaceutical and food products

Tse-Chao Hua, Bao-Lin Liu and Hua Zhang

CRC Press

Boca Raton Boston New York Washington, D. C.

WOODHEAD PUBLISHING LIMITED

Oxford Cambridge New Delhi

科学出版社
Science Press

Published by Woodhead Publishing Limited, Abington Hall, Granta Park, Great Abington, Cambridge CB21 6AH, UK
www.woodheadpublishing.com

Woodhead Publishing India Private Limited, G-2, Vardaan House, 7/28 Ansari Road, Daryaganj, New Delhi -- 110002, India
www.woodheadpublishingindia.com

Published in China by Science Press, 16 Donghuangchenggen North Street, Beijing 100717, China

Published in North America by CRC Press LLC, 6000 Broken Sound Parkway, NW, Suite 300, Boca Raton, FL 33487, USA

First published 2010, Woodhead Publishing Limited, Science Press and CRC Press LLC

British Library Cataloguing in Publication Data
A catalogue record for this book is available from the British Library.

Library of Congress Cataloging in Publication Data
A catalog record for this book is available from the Library of Congress

Woodhead Publishing ISBN 978-0-08-101464-6 (book)
Woodhead Publishing ISBN 978-1-84569-747-1 (e-book)
CRC Press ISBN 978-1-4398-2598-3
CRC Press order number N10143

Typeset by Beijing Easthuman Co. Ltd., Beijing, China
Printed by Shuang Qing Printing House, Beijing, China

Contents

Woodhead Publishing Series in Food Science, Technology and Nutrition

1 **Chilled foods: a comprehensive guide** *Edited by C. Dennis and M. Stringer*
2 **Yoghurt: science and technology** *A. Y. Tamime and R. K. Robinson*
3 **Food processing technology: principles and practice** *P. J. Fellows*
4 **Bender's dictionary of nutrition and food technology Sixth edition** *D. A. Bender*
5 **Determination of veterinary residues in food** *Edited by N. T. Crosby*
6 **Food contaminants: sources and surveillance** *Edited by C. Creaser and R. Purchase*
7 **Nitrates and nitrites in food and water** *Edited by M. J. Hill*
8 **Pesticide chemistry and bioscience: the food-environment challenge** *Edited by G. T. Brooks and T. Roberts*
9 **Pesticides: developments, impacts and controls** *Edited by G. A. Best and A. D. Ruthven*
10 **Dietary fibre: chemical and biological aspects** *Edited by D. A. T. Southgate, K. W. Waldron, I. T. Johnson and G. R. Fenwick*
11 **Vitamins and minerals in health and nutrition** *M. Tolonen*
12 **Technology of biscuits, crackers and cookies Second edition** *D. Manley*
13 **Instrumentation and sensors for the food industry** *Edited by E. Kress-Rogers*
14 **Food and cancer prevention: chemical and biological aspects** *Edited by K. W. Waldron, I. T. Johnson and G. R. Fenwick*
15 **Food colloids: proteins, lipids and polysaccharides** *Edited by E. Dickinson and B. Bergenstahl*

16 **Food emulsions and foams** *Edited by E. Dickinson*
17 **Maillard reactions in chemistry, food and health** *Edited by T. P. Labuza, V. Monnier, J. Baynes and J. O'Brien*
18 **The Maillard reaction in foods and medicine** *Edited by J. O'Brien, H. E. Nursten, M. J. Crabbe and J. M. Ames*
19 **Encapsulation and controlled release** *Edited by D. R. Karsa and R. A. Stephenson*
20 **Flavours and fragrances** *Edited by A. D. Swift*
21 **Feta and related cheeses** *Edited by A. Y. Tamime and R. K. Robinson*
22 **Biochemistry of milk products** *Edited by A. T. Andrews and J. R. Varley*
23 **Physical properties of foods and food processing systems** *M. J. Lewis*
24 **Food irradiation: a reference guide** *V. M. Wilkinson and G. Gould*
25 **Kent's technology of cereals: an introduction for students of food science and agriculture Fourth edition** *N. L. Kent and A. D. Evers*
26 **Biosensors for food analysis** *Edited by A. O. Scott*
27 **Separation processes in the food and biotechnology industries: principles and applications** *Edited by A. S. Grandison and M. J. Lewis*
28 **Handbook of indices of food quality and authenticity** *R. S. Singhal, P. K. Kulkarni and D. V. Rege*
29 **Principles and practices for the safe processing of foods** *D. A. Shapton and N. F. Shapton*
30 **Biscuit, cookie and cracker manufacturing manuals Volume 1: ingredients** *D. Manley*
31 **Biscuit, cookie and cracker manufacturing manuals Volume 2: biscuit doughs** *D. Manley*
32 **Biscuit, cookie and cracker manufacturing manuals Volume 3: biscuit dough piece forming** *D. Manley*
33 **Biscuit, cookie and cracker manufacturing manuals Volume 4: baking and cooling of biscuits** *D. Manley*
34 **Biscuit, cookie and cracker manufacturing manuals Volume 5: secondary processing in biscuit manufacturing** *D. Manley*
35 **Biscuit, cookie and cracker manufacturing manuals Volume 6: biscuit packaging and storage** *D. Manley*
36 **Practical dehydration Second edition** *M. Greensmith*
37 **Lawrie's meat science Sixth edition** *R. A. Lawrie*
38 **Yoghurt: science and technology Second edition** *A. Y. Tamime and R. K. Robinson*
39 **New ingredients in food processing: biochemistry and agriculture** *G. Linden and D. Lorient*
40 **Benders' dictionary of nutrition and food technology Seventh edition** *D. A. Bender and A. E. Bender*
41 **Technology of biscuits, crackers and cookies Third edition** *D. Manley*
42 **Food processing technology: principles and practice Second edition** *P. J. Fellows*

43 **Managing frozen foods** *Edited by C. J. Kennedy*
44 **Handbook of hydrocolloids** *Edited by G. O. Phillips and P. A. Williams*
45 **Food labelling** *Edited by J. R. Blanchfield*
46 **Cereal biotechnology** *Edited by P. C. Morris and J. H. Bryce*
47 **Food intolerance and the food industry** *Edited by T. Dean*
48 **The stability and shelf life of food** *Edited by D. Kilcast and P. Subramaniam*
49 **Functional foods: concept to product** *Edited by G. R. Gibson and C. M. Williams*
50 **Chilled foods: a comprehensive guide Second edition** *Edited by M. Stringer and C. Dennis*
51 **HACCP in the meat industry** *Edited by M. Brown*
52 **Biscuit, cracker and cookie recipes for the food industry** *D. Manley*
53 **Cereals processing technology** *Edited by G. Owens*
54 **Baking problems solved** *S. P. Cauvain and L. S. Young*
55 **Thermal technologies in food processing** *Edited by P. Richardson*
56 **Frying: improving quality** *Edited by J. B. Rossell*
57 **Food chemical safety Volume 1: contaminants** *Edited by D. Watson*
58 **Making the most of HACCP: learning from others' experience** *Edited by T. Mayes and S. Mortimore*
59 **Food process modelling** *Edited by L. M. M. Tijskens, M. L. A. T. M. Hertog and B. M. Nicolaï*
60 **EU food law: a practical guide** *Edited by K. Goodburn*
61 **Extrusion cooking: technologies and applications** *Edited by R. Guy*
62 **Auditing in the food industry: from safety and quality to environmental and other audits** *Edited by M. Dillon and C. Griffith*
63 **Handbook of herbs and spices Volume 1** *Edited by K. V. Peter*
64 **Food product development: maximising success** *M. Earle, R. Earle and A. Anderson*
65 **Instrumentation and sensors for the food industry Second edition** *Edited by E. Kress-Rogers and C. J. B. Brimelow*
66 **Food chemical safety Volume 2: additives** *Edited by D. Watson*
67 **Fruit and vegetable biotechnology** *Edited by V. Valpuesta*
68 **Foodborne pathogens: hazards, risk analysis and control** *Edited by C. de W. Blackburn and P. J. McClure*
69 **Meat refrigeration** *S. J. James and C. James*
70 **Lockhart and Wiseman's crop husbandry Eighth edition** *H. J. S. Finch, A. M. Samuel and G. P. F. Lane*
71 **Safety and quality issues in fish processing** *Edited by H. A. Bremner*
72 **Minimal processing technologies in the food industries** *Edited by T. Ohlsson and N. Bengtsson*
73 **Fruit and vegetable processing: improving quality** *Edited by W. Jongen*
74 **The nutrition handbook for food processors** *Edited by C. J. K. Henry and C. Chapman*

107 **Functional foods, cardiovascular disease and diabetes** *Edited by A. Arnoldi*
108 **Brewing: science and practice** *D. E. Briggs, P. A. Brookes, R. Stevens and C. A. Boulton*
109 **Using cereal science and technology for the benefit of consumers: proceedings of the 12th International ICC Cereal and Bread Congress, 24–26th May, 2004, Harrogate, UK** *Edited by S. P. Cauvain, L. S. Young and S. Salmon*
110 **Improving the safety of fresh meat** *Edited by J. Sofos*
111 **Understanding pathogen behaviour in food: virulence, stress response and resistance** *Edited by M. Griffiths*
112 **The microwave processing of foods** *Edited by H. Schubert and M. Regier*
113 **Food safety control in the poultry industry** *Edited by G. Mead*
114 **Improving the safety of fresh fruit and vegetables** *Edited by W. Jongen*
115 **Food, diet and obesity** *Edited by D. Mela*
116 **Handbook of hygiene control in the food industry** *Edited by H. L. M. Lelieveld, M. A. Mostert and J. Holah*
117 **Detecting allergens in food** *Edited by S. Koppelman and S. Hefle*
118 **Improving the fat content of foods** *Edited by C. Williams and J. Buttriss*
119 **Improving traceability in food processing and distribution** *Edited by I. Smith and A. Furness*
120 **Flavour in food** *Edited by A. Voilley and P. Etievant*
121 **The Chorleywood bread process** *S. P. Cauvain and L. S. Young*
122 **Food spoilage microorganisms** *Edited by C. de W. Blackburn*
123 **Emerging foodborne pathogens** *Edited by Y. Motarjemi and M. Adams*
124 **Benders' dictionary of nutrition and food technology Eighth edition** *D. A. Bender*
125 **Optimising sweet taste in foods** *Edited by W. J. Spillane*
126 **Brewing: new technologies** *Edited by C. Bamforth*
127 **Handbook of herbs and spices Volume 3** *Edited by K. V. Peter*
128 **Lawrie's meat science Seventh edition** *R. A. Lawrie in collaboration with D. A. Ledward*
129 **Modifying lipids for use in food** *Edited by F. Gunstone*
130 **Meat products handbook: practical science and technology** *G. Feiner*
131 **Food consumption and disease risk: consumer-pathogen interactions** *Edited by M. Potter*
132 **Acrylamide and other hazardous compounds in heat-treated foods** *Edited by K. Skog and J. Alexander*
133 **Managing allergens in food** *Edited by C. Mills, H. Wichers and K. Hoffman-Sommergruber*
134 **Microbiological analysis of red meat, poultry and eggs** *Edited by G. Mead*
135 **Maximising the value of marine by-products** *Edited by F. Shahidi*
136 **Chemical migration and food contact materials** *Edited by K. Barnes, R. Sinclair and D. Watson*

137 **Understanding consumers of food products** *Edited by L. Frewer and H. van Trijp*

138 **Reducing salt in foods: practical strategies** *Edited by D. Kilcast and F. Angus*

139 **Modelling microrganisms in food** *Edited by S. Brul, S. Van Gerwen and M. Zwietering*

140 **Tamime and Robinson's Yoghurt: science and technology Third edition** *A. Y. Tamime and R. K. Robinson*

141 **Handbook of waste management and co-product recovery in food processing: Volume 1** *Edited by K. W. Waldron*

142 **Improving the flavour of cheese** *Edited by B. Weimer*

143 **Novel food ingredients for weight control** *Edited by C. J. K. Henry*

144 **Consumer-led food product development** *Edited by H. MacFie*

145 **Functional dairy products Volume 2** *Edited by M. Saarela*

146 **Modifying flavour in food** *Edited by A. J. Taylor and J. Hort*

147 **Cheese problems solved** *Edited by P. L. H. McSweeney*

148 **Handbook of organic food safety and quality** *Edited by J. Cooper, C. Leifert and U. Niggli*

149 **Understanding and controlling the microstructure of complex foods** *Edited by D. J. McClements*

150 **Novel enzyme technology for food applications** *Edited by R. Rastall*

151 **Food preservation by pulsed electric fields: from research to application** *Edited by H. L. M. Lelieveld and S. W. H. de Haan*

152 **Technology of functional cereal products** *Edited by B. R. Hamaker*

153 **Case studies in food product development** *Edited by M. Earle and R. Earle*

154 **Delivery and controlled release of bioactives in foods and nutraceuticals** *Edited by N. Garti*

155 **Fruit and vegetable flavour: recent advances and future prospects** *Edited by B. Brückner and S. G. Wyllie*

156 **Food fortification and supplementation: technological, safety and regulatory aspects** *Edited by P. Berry Ottaway*

157 **Improving the health-promoting properties of fruit and vegetable products** *Edited by F. A. Tomás-Barberán and M. I. Gil*

158 **Improving seafood products for the consumer** *Edited by T. Børresen*

159 **In-pack processed foods: improving quality** *Edited by P. Richardson*

160 **Handbook of water and energy management in food processing** *Edited by J. Klemeš, R. Smith and J.-K. Kim*

161 **Environmentally compatible food packaging** *Edited by E. Chiellini*

162 **Improving farmed fish quality and safety** *Edited by Ø. Lie*

163 **Carbohydrate-active enzymes** *Edited by K.-H. Park*

164 **Chilled foods: a comprehensive guide Third edition** *Edited by M. Brown*

165 **Food for the ageing population** *Edited by M. M. Raats, C. P. G. M. de Groot and W. A. Van Staveren*

166 **Improving the sensory and nutritional quality of fresh meat** *Edited by J. P. Kerry and D. A. Ledward*

167 **Shellfish safety and quality** *Edited by S. E. Shumway and G. E. Rodrick*

168 **Functional and speciality beverage technology** *Edited by P. Paquin*

169 **Functional foods: principles and technology** *M. Guo*

170 **Endocrine-disrupting chemicals in food** *Edited by I. Shaw*

171 **Meals in science and practice: interdisciplinary research and business applications** *Edited by H. L. Meiselman*

172 **Food constituents and oral health: current status and future prospects** *Edited by M. Wilson*

173 **Handbook of hydrocolloids Second edition** *Edited by G. O. Phillips and P. A. Williams*

174 **Food processing technology: principles and practice Third edition** *P. J. Fellows*

175 **Science and technology of enrobed and filled chocolate, confectionery and bakery products** *Edited by G. Talbot*

176 **Foodborne pathogens: hazards, risk analysis and control Second edition** *Edited by C. de W. Blackburn and P. J. McClure*

177 **Designing functional foods: measuring and controlling food structure breakdown and absorption** *Edited by D. J. McClements and E. A. Decker*

178 **New technologies in aquaculture: improving production efficiency, quality and environmental management** *Edited by G. Burnell and G. Allan*

179 **More baking problems solved** *S. P. Cauvain and L. S. Young*

180 **Soft drink and fruit juice problems solved** *P. Ashurst and R. Hargitt*

181 **Biofilms in the food and beverage industries** *Edited by P. M. Fratamico, B. A. Annous and N. W. Gunther*

182 **Dairy-derived ingredients: food and neutraceutical uses** *Edited by M. Corredig*

183 **Handbook of waste management and co-product recovery in food processing Volume 2** *Edited by K. W. Waldron*

184 **Innovations in food labelling** *Edited by J. Albert*

185 **Delivering performance in food supply chains** *Edited by C. Mena and G. Stevens*

186 **Chemical deterioration and physical instability of food and beverages** *Edited by L. Skibsted, J. Risbo and M. Andersen*

187 **Managing wine quality Volume 1: viticulture and wine quality** *Edited by A. Reynolds*

188 **Improving the safety and quality of milk Volume 1: milk production and processing** *Edited by M. Griffiths*

189 **Improving the safety and quality of milk Volume 2: improving quality in milk products** *Edited by M. Griffiths*

190 **Cereal grains: assessing and managing quality** *Edited by C. Wrigley and I. Batey*

191 **Sensory analysis for food and beverage control: a practical guide** *Edited by D. Kilcast*

192 **Managing wine quality Volume 2: oenology and wine quality** *Edited by A. Reynolds*

193 **Winemaking problems solved** *Edited by C. Butzke*

194 **Environmental assessment and management in the food industry** *Edited by U. Sonneson, J. Berlin and F. Ziegler*

195 **Consumer-driven innovation in food and personal products** *Edited by H. MacFie and S. Jaeger*

196 **Tracking pathogens in the food chain** *Edited by S. Brul, P. M. Fratamico and T. A. McMeekin*

197 **Case studies in novel food processing technologies** *Edited by C. Doona, K. Kustin and F. Feeherry*

198 **Freeze-drying of pharmaceutical and food products** *Tse-Chao Hua, Bao-Lin Liu and Hua Zhang*

Preface

Freeze-drying is a kind of preservation technology, by which the material is cooled below its eutectic temperature or glass transition temperature firstly to be solidified completely, then dried in vacuum space at low temperature by sublimation drying and desorption drying till 95%—99% of moisture is removed. The product can be stored at room temperature or 4 ℃ for a long time. Freeze-drying has become a most important technique for the preservation of heat-sensitive pharmaceuticals and foods. Meanwhile, the process of freeze drying may greatly affect the quality of the products.

Freeze-drying technology has found many applications, such as preservation of foods, micro-organisms, biological pharmaceuticals, human cells, and preparation of various superfine powders. Freeze-dried food has marked advantages: keeping fresh color, smell and taste; avoiding loss of nutritional ingredients and surface hardening. It is light in weight and easy to rehydrate; it is a high value-added product. Freeze drying of biological pharmaceuticals has become the most important application in the freeze drying industry in the past 10 years, and it is a field of most concern and with largest investment in freeze-drying. The pharmaceutical products made by freeze drying have the following characteristics: structural stability, basically unchanged biological activity, almost no loss of volatile ingredients and heat sensitive ingredients, porous structure and good therapeutic effectiveness, high water loss of 95%—99%, convenient storage at room temperature or in the refrigerator for long time.

Freeze-drying has become a most important technique for the preservation of heat-sensitive pharmaceuticals and foods. Meanwhile, the process of freeze drying may greatly affect the quality of the products.

Although the method of freeze drying was known quite early, it was still in its primary "rough" stage till the 1990s. Back to the 1990s, the appearance and development of biological pharmaceutical products proposed many "harsh" demands to the freeze-drying technology, forcing it developing to its "delicate" stage. At the same time, the "Solution vitrification theory"and "Food polymer science theory" had provided some theoretical foundation to the development of freeze drying technology.

The authors have been involved the field of cryopreservation since 1980, and doing some researches in freeze drying of foods, pharmaceuticals and human cells from 1996 on. A Chinese book with the same title was edited and published by Science Publisher in 2006.

The book contains 8 chapters. They are introduction, fundamentals, heat-mass transfer analyses and modeling, equipment, freeze-drying of food, protective agent and additives, freeze-drying of pharmaceuticals, disinfection sterilization and valuation. Since the temperature history of freeze and drying processes has great effects on the quality of product and the time & energy consumption, more attention is paid in this book to the mechanism and the parameters controlling of freeze drying process.

In this English edition, translation and some amendments have been made by the authors, Chapters 1—3 by Hua T C; Chapter 4 by Zhang H; Chapters 5—8 by Liu B L. Our colleague Hua K F has made improvement in English; and graduate students, Li M, Shi M J, Zhou Y, Zhou N, Zhou C M, Yang M have taken part some work. The authors wish to acknowledge their assistance. The authors wish to acknowledge the supports of projects (S30503, NCET-07-0559, NSFC and USST).

The authors
March, 2009
in Shanghai University of Science and Technology
(http://www.usst.edu.cn)

1

Introduction

1.1 Brief history of freeze-drying technology

When washed clothes are put outside in the sun in severely cold winter with temperature below 0℃, they will be frozen very quickly; however through a period of time, they will be dried, since the moisture frozen in the clothes is sublimated to the air. The drier the air, or the lower the partial pressure of vapor in the atmosphere, the quicker the sublimation drying goes.

The method of freeze-drying food was known a little to the ancient Chinese and Peruvian Incas. In order to store meat longer and to get better flavor, the ancient Chinese put the meat outside in cold winter. The meat was firstly frozen, and then dried. The ancient Incas stored potatoes and other some crops on the mountain heights above Machu Picchu. The cold mountain temperature froze the food and the water inside slowly sublimated under the low pressure of the high altitudes. Freeze-dried food is light and can last longer than other preserved food.

The above phenomena may be taken as examples of "freeze-drying". However as a part of science and technology, "freeze-drying" is only a matter of recent 80 years.

There were three events which had the milestone significance in the early development in freeze-drying technology:

(1) Earl W. Flosdorf and Stuart Mudd from University of Pennsylvania successfully preserved the human serum by freeze-drying using glass apparatus in 1933[1].

(2) During World War II, the freeze-drying process was developed commercially when it was used to preserve blood plasma and penicillin. Howard Walter Florey and Ernst Boris Chain, the scientists who followed up most successfully on Alexander Fleming's discovery of penicillin in 1928, devised the method to store the penicillin by freeze-drying in 1938, and made

great contribution to clinical application during the World War II. The three scientists were awarded the Nobel Prize for Physiology or Medicine in 1945. Several years later, Charles Merieux (French virologist) made the freeze drying as a regular technology for vaccine preservation.

(3) In 1930, Nestlé's chairman was approached by the Brazilian Coffee Institute and asked to develop coffee that was soluble in hot water and retaining its flavor. At the time, Brazil had a huge coffee surplus. After seven years of painstaking research at the Nestlé research centre laboratory in Switzerland, scientist Max Mortgenthaler invented freeze-dried coffee and the powdered coffee was first produced in 1938, which led to the development of powdered food products.

These three events impelled the freeze-drying development of the microorganism and food. However the system and apparatus used for freeze drying are not convenient and efficient at that time. Most apparatus are made of fragile glass. The necessary vacuum is produced by a "chemical pump" with ethyl ether, not mechanical vacuum pump. The low temperatures making the materials to be frozen are reached by applying "dry ice" (solid CO_2).

Freeze-drying is a method of preservation by refrigeration, heating and vacuum. Firstly the materials are cooled to low temperature, and the water in the material is almost completely frozen. Then the ice inside the materials is heated and sublimated from the solid state directly into the vapor phase during sublimation drying process. The freeze dried materials can be stored at normal temperature in vacuum state. Thus the development of freeze-drying depends on the improvement of refrigeration and vacuum technology.

The mechanical refrigeration based on vapor compression and throttling decreasing temperature was rapidly developed in 1930s. DuPont Company in USA produced a new commercial refrigerant of R-12, trademarked as Freon, in 1931. The rise of mechanical vapor compression refrigeration in 1930s, had greatly promoted the development and application of freeze-drying technology. The technology had its rapid development period in the 1940—1960s, and had main uses in microorganism, coffee and so on in those years.

After the first commercial freeze-dryer came out in 1935, the technology started from the laboratory stage to the industrial production and the product commercialization stage. Under the vigorous impetuses by Flosdorf, Greaves, Henaff and the many others, freezing and drying became a conventional method of preserving blood plasma and the application in the Second World War had saved millions human's life.

Food freeze-drying studies started in Flosdorf's laboratory in the 1930s. The Britisch Food Department had also conducted the research of food freeze drying in the Scotland Aberdeen pilot plant.

Freeze-drying has become one of the most important processes for the preservation of heat-sensitive pharmaceuticals and foods. Recently the freeze-drying of human living cells is in study[5].

Along with the promotion of freezing and drying application, theoretical and technological research also prospered in a steady pace. Flosdorf published the first monograph of freeze-drying in 1944. The first and second symposiums on freeze-drying were held in London in 1951 and 1958 respectively. American was the first in the world to establish the GMP (Good Manufacturing Practice) production standard for the freeze-drying of pharmaceutical products in 1963. Many nations also established the production standard afterwards, and a resolution to execute GMP on freeze-drying of pharmaceutical products was passed in WTO in 1969.

Hereafter there was a long period of time during which the freeze-drying technology had little development. Since there were no many demands, the technology was in its primary "rough" stage. As far as to the 1990s, the appearance and development of biological pharmaceutical products proposed many "harsh" demands to the freeze-drying technology, forcing it developing to "delicate" stage. At the same time,"Solution vitrification theory" and "Food polymer science theory" had provided some theoretical foundation to the development of freeze-drying.

Application of freeze-drying in material science is a matter of the recent years. Nanometer technology has risen abruptly on the horizon. Freeze drying has become one of the most important manufacture technology of nanoparticles.

1.2 Basic processes of freeze-drying

Freeze-drying is also called "lyophilization". The term "lyophilization" was derived from the word "*lyophile*", coming from the Greek λυοζ and φιλειν, which means "likes the solvent" (the solvent is water in most cases), describing the great ability of the dry product to rehydrate again[2].

The freeze drying is constituted of the following processes[2—4].

1.2.1 Preparation or pretreatment of the material

For the food materials, it is necessary to do some physical and chemical pretreatments before freeze drying, which includes cleanout, classification, slice up, blanching, sterilization, concentration and others. There are differences for different kinds of food. In most cases, it is not necessary to add additive to the food material.

For the pharmaceutical materials or the living cells, it is necessary to add some additives before freeze-drying in order to keep their activity and get good quality of final products. It is called 'formulation' for mixture of pharmaceutical material and additives. The additives are used for stabilizing the formulation or for therapeutic reasons, which can be classified by their functions as lyoprotectant (protectant agent for freeze drying), emulsifier, bulking agent, antioxidant, buffer agent and others. Recently, it is found that some kinds of sugar, such as trehalosedihydrate and sucrose, are efficient lyoprotectants.

1.2.2 Cooling and solidification of the material

The cooling process should be done sufficiently, i.e., not only the free water in the material should be frozen to form crystalline ice completely, but also the other constitutes of the material should be solidified, to form non-crystalline solid (glassy solid). This process is often called a freezing process, actually it is a completed solidification process, while the material should be completely solidified by enough cooling. The solidified material after perfect cooling contains both crystalline solid and non-crystalline hard reticulation structure solid.

The cooling and solidification process is an extremely important process, however it is underestimated in the past. For a perfect cooling process, the final temperature, the temperature of complete solidification, T_{cs}, should be lower than the eutectic temperature, T_e, or the glass transition temperature T_g.

The temperature decreasing rate of the cooling and solidification process is also an important effect. In general, rapid cooling would lead to formation of partial glassy state and prevent excessive dehydration of material during cooling. However, too speedy cooling would cause harmful results such as damaging the living cells and fracture of the material.

The temperature variation with time is called "thermal history". The thermal history during the cooling process is also very important because it will influence the thermal properties of the material, which is similar to the heat treatment process of metal. It has been found that annealing for a length of time would cause some properties variation, such as glass transition temperature T_g.

1.2.3 Sublimation drying (primary drying)

For the analysis of freeze-drying, the water inside the material can be classified into two kinds. One is freezable water in low temperature, which is also called free water; the other kind can not be frozen at low temperature. This part of water is bound, therefore it is called "bound water". Generally it is thought the bound water is adsorbed by the material. For the materials rich of water, the free water will reach over 90% of the total moisture content.

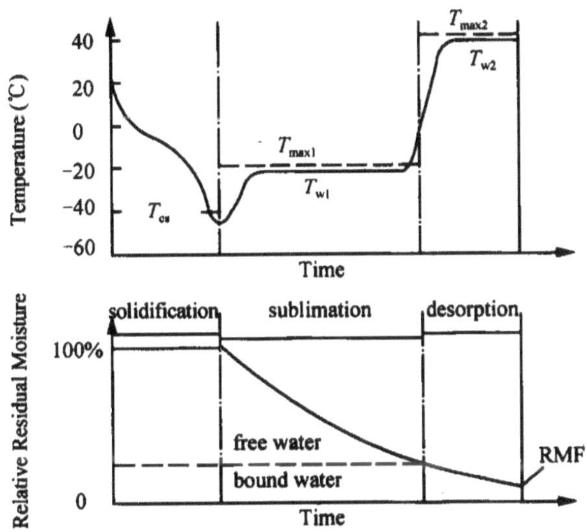

Fig.1.1 The variations of material temperature and moisture during freeze-drying processes

Fig.1.1 indicates the variations of the material temperature and the moisture during freeze-drying processes. The abscissa denotes the time; and the ordinate denotes temperature and relative moisture content for the upper part chart and the lower part chart respectively. The relative water content is 100% initially; and finally it is RMF, namely the requested residual moisture final.

Sublimation drying, so called primary drying, refers to heating the material at low temperature to cause frozen "free water" to sublimate directly to vapor. The material temperature during sublimation must be lower than the material highest permissible temperature T_{max1}. The highest permissible temperature T_{max1} may be the glass transition temperature T_g, or the eutectic temperature T_e. If the sublimation temperature is too high, the material will appear softened or collapsed.

During primary drying, the quantity of heat needed is the heat of ice sublimation. The heating can be done by heat conduction or by heat radiation. Two basic conditions must be satisfied to assure the progress of the sublimation process: the vapor sublimated must be moved away unceasingly from the sublimation interface; and the necessary heat must be unceasingly provided to the material for sublimation. If anyone of the two basic conditions is not met, some phenomena would appear, such as softening, thawing, bulging or collapsing. In fact the sublimation drying is a process in which the heat transfer and the mass transfer simultaneously carry on. Only when the heat rate transferred to the interface of sublimation is equal to the heat rate needed for the vapor escaped from the interface, the sublimation drying can carry on smoothly. The heat transfer and mass transfer of the material are subject to many restrictions, therefore the sublimation drying is a very time-consuming process.

1.2.4 Desorption drying (secondary drying)

After primary drying, there is still unfrozen water adsorbed on porous structure surface and on the polar groups inside the dried material. As the adsorption energy is quite high, desorption the bound water also needs higher temperature and enough quantity of heat. However the temperature of desorption shall not be too high, otherwise it would cause denaturation of the materials.

Desorption drying, also known as the secondary drying, is heating at a higher temperature, so that the bound unfrozen water absorbs the heat of desorption and becomes free water; then the free water absorbs the heat of evaporation and becomes vapor escaping from the material finally. During the desorption drying process, the heating temperature of the material T_{w2} must be less than the maximum allowed temperature T_{max2}. The maximum permissible temperature T_{max2} is decided by the material nature. For protein pharmaceuticals, the maximum allowed temperature should generally be lower than 40℃; for food, such as fruits and vegetables, the maximum permissible temperature may arrive 60—70℃ or higher.

In the secondary drying process, the heat needed is the sum of desorption heat and evaporation heat, and it is simply called as "the desorption heat". At the end of desorption drying process, the moisture left inside the material is required to reach the remaining moisture final RMF. Too high or too low RMF of the freeze-dried material is detrimental. Too high RMF is not conducive to long-term storage; too low RMF will harm the active material. Generally the remaining moisture final RMF should be less than 5%.

1.2.5 Conditioning-packing and storage

After primary and secondary stages of freeze drying, the products must go through conditioning packing and storage. Sealing and packing should be conducted in dry environment. If no contacting with oxygen or water vapor in the air, the freeze dried products may store longer. When the product is needed to use, rehydrating it is necessary in most cases.

The packing and sealing should be carried out either in the vacuum chamber or in a chamber with sufficient inert gas (nitrogen or argon). For freeze dried material in bottle or vial, sealing may be carried out in the drying chamber, with rubber stopper pressed to stop up the vapor channel, as shown in Fig. 1.2. For bulk products or material in ampoules, they can be drawn out from the drying chamber through a vacuum channel, delivered to a vacuum chamber (or a chamber with sufficient inert gas) and sealed by the manipulator.

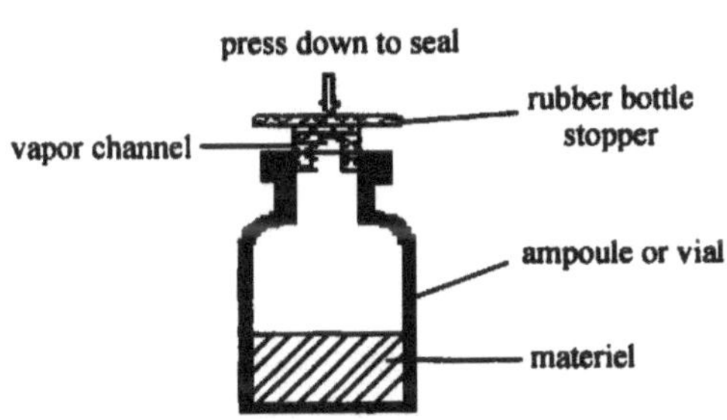

Fig.1.2 A vapor channel for drying and a stopper for sealing of a bottle or vial

Generally, the storage of freeze dried products should be conduct at room temperature. Regarding certain drugs, the storage temperature of 4 ℃ is required; and for some special products −18 ℃ is required. All these requirements can be easily satisfied by ordinary refrigerator.

1.3 Protecting agents of freeze-drying

During freeze-drying process and in the storage period of foods, drugs and organisms, many factors, e.g., chemical composition, cooling rate , freeze and dehydrated stress, glass transition temperature, residual moisture content, temperature, humidity of the storage environment and so on, can affect the stability of active constituent or even cause the loss of activation. The massive experimental study indicated that except minority materials, such as some foods, the human blood plasma, the milk, may directly carry on the freeze-drying process; the majority of drugs and the biological preparations need addition of some appropriate protecting agents and some chemical additive for freeze-drying and storage.

So far there are several classifications of protecting agents as below[2–4].

1.3.1 Classification of protecting agents by their molecular weight

1. low-molecular weight compounds

They may also be classified as the acidic, the neutral and the alkaline. Acidic materials include glutanic acid, aspartic acid, apple ammonia acid, lactic acid and so on. Neutral materials include glucose, phaseomannite, lactose, sucrose, raffinose, trohalose, sorbitol DL- threonine, phaseomannite, xylitol and so on. Alkaline materials include arginine and histidine and so on.

2. high-molecular weight compounds

They are such as albumin, gelatin, protein peptone, soluble starch, dextrin, meat broth, pectin, Arabic gum, methylol cellulose, algae and some natural mixtures like degreasing milk, blood serum and so on.

As widely acknowledged, functional mechanism of protecting agent is that the low molecular compounds function directly on freezing and drying process while the high molecular compounds promote the protective function of low molecular compound. Therefore, when preparing a protecting agent formula, low-molecular and high-molecular compounds are used together.

1.3.2 Classification of protecting agents by their function

1. lyoprotectant

Lyoprotectant, namely freeze drying protecting agent, is a kind of substance which can prevent the active constituent from denaturation during the freezing and drying processes, such as glycerine, DMSO, trohalose, sucrose, polyvinylpyrrolidone (PVP) and so on.

2. bulking agent (bulking compound)

Bulking agent, namely bulking compound, is a kind of substance which can prevent the effective components of the formula from escaping along with the water vapor, and can promote the fixation of the effective components in the material, such as mannitol, lactose, gelatin and so on.

3. antioxidant

Antioxidant, namely oxidation inhibitor, is a kind of substance which can prevent the effective components of the material from oxidation and denaturation during freeze drying and storage, such as Vitamin D, Vitamin E, protein hydrolysate, sodium hyposulfite and so on.

4. buffer agent (PH modifier)

Buffer agent, also called PH modifier, is a kind of substance which can regulate the acid and alkali value of the material to the most stable region for active material, such as phosphoric acid, sorbitol, EDTA (ethylenediamine tetraacetic acid), amino acid and so on.

1.3.3 Classification of protecting agents by their material kinds

They are sugars/polysaccharides polymers, surfactants, amino acids, salts, etc.

In the formula of freeze-drying material, besides the active constituent and the solvent, many other kinds of chemical additives must also be used. These chemical additives do not have a common name until now. They may be called protecting agents, chemical additive, or sometimes excipient. The word "excipient" originated from pharmaceutics. It referred to non-active supplementary material (for example Arabic gum, syrup, starch), especially the supplementary material added to those medicine mixtures with massive liquid in order to prepare pills and tablets of medicine. However "excipient" has been greatly expanded to refer to all kinds of chemical additives up to now. According to literature statistics, "excipient" is used quite frequently in freeze-drying formula.

In freeze-drying formula, there are some chemical additives playing one specific role and other chemical additives play several roles simultaneously. For example PVP (Polyvinylpyrrolidone) may serve as the low temperature protecting agent and bulking agent at the same time.

The specific functions of a chemical additive in the formula are difficult to identify strictly. Sometimes an identical additive may display quite different functions in several different freeze dried products.

1.4 Freeze-drying system constitution

1.4.1 Requirement of ice sublimation to freeze-drying system

The trip point of water is at 0.01 ℃ and 610.62 Pa. The saturated pressure of ice reduces along with temperature drop, as shown in Table 1.1.

Table 1.1 The saturated pressure of ice with temperature

Temperature (℃)	-70	-60	-50	-40	-30	-20	-10	0.01
Saturated pressure (Pa)	0.262	1.11	3.94	12.9	38.0	103.3	259.9	610.62

There are two ways to sublimate the ice to vapor as shown in Fig.1.3. One is by heating to elevate the temperature; another is by evacuating to reduce pressure. In fact, those two methods are simultaneously applied.

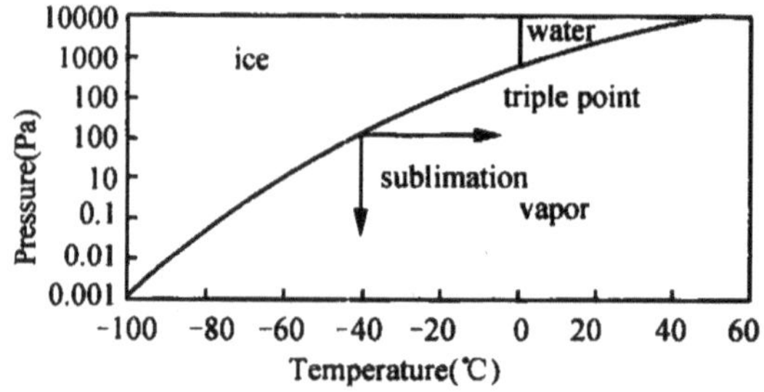

Fig. 1.3 The sublimation process of ice

For example, in order to make the ice to sublimate at −40℃, the pressure inside the drying chamber must be lower than 12.9Pa. At the same time in order to provide the heat which the ice sublimation needs, material must be heated. The sublimation heat of ice at −40℃ is 2838.6kJ/kg.

1.4.2 Main components of freeze-drying system[2–7]

A freeze-drying system is mainly composed of the following components: a drying chamber, a cold trap, a refrigeration system, a vacuum system, a heating system and a control system, etc., as shown in Fig. 1.4.

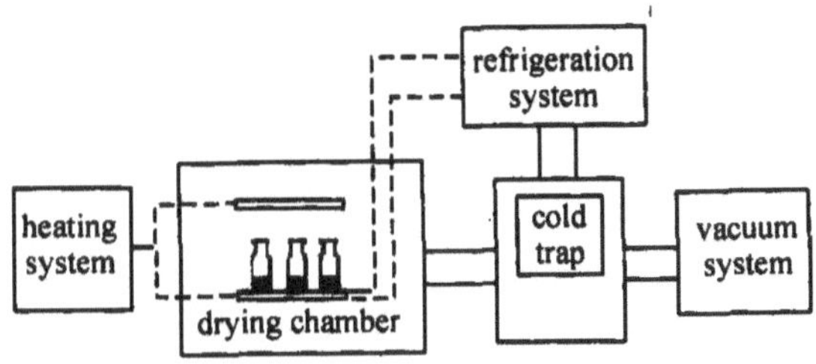

Fig.1.4 The composition of a freeze-drying system

1.4.3 Technical requirement of major component

(1) The drying chamber is the place in which drying process under vacuum carries on. The bottles or vials with materials inside are put on the lower heating plate; there is upper heating plate sometimes, as shown in Fig.1.4. Heating materials can be achieved by the conduction through the lower plate, or by the radiation from the upper plate.

In the freeze drying process of pharmaceutical materials, it is required that all the drying chamber and the vacuum system are aseptic, therefore how to keep clean and antiseptic is always considered. Generally high temperature steam is used for sterilizing, therefore the drying chamber is made of stainless steel, moreover the entire accessories inside the chamber should be able to tolerate high temperature steam.

(2) The cooling solidification process (or the freezing process) of material may be performed outside the drying chamber. After the material being cooled completely, it is put into the drying chamber. The cooling process may also be performed inside the drying chamber. This time there is a refrigerant or secondary refrigerant flowing inside the lower plate, refrigeration system worked to decrease the temperature of the material.

(3) The drying chamber is requested to be well sealed. In some drying chambers, there is a manipulator which can press the rubber stoppers to seal the bottles.

(4) Cold trap is a very important component in the freeze-drying system. The function of cold trap is to provide a low-temperature environment in a vacuum system. The temperature of the cold trap should be much lower than the sublimation interface temperature of the material, 20K or more. In this way, the saturated vapor pressure near the cold trap would be much lower than that of the sublimation interface of the material. The pressure drop is the driving force of mass transfer which presses the water vapor flowing from the interface of the material to the clod trap.

The water vapor escaped from the material is condensed into liquid when it contacts with the cold surface of the cold trap, and then is expelled from the system. Therefore the cold trap can be considered as a condenser of water vapor in the freeze drying system.

The cold trap can be installed inside or outside the drying chamber. The later style is shown in Fig.1.4.

For sublimation drying, if the material sublimation temperature is around -10℃ to -40℃, the temperature of cold trap should be in -30℃ to -70℃. The larger the temperature difference between the materials sublimation temperature and the cold trap, the greater the driving force of the mass transfer. However, too large temperature difference between the two is meaningless. For example, the material sublimation temperature is -40℃, corresponding to its saturation vapor pressure of water vapor 12.9 Pa; cold trap temperature of -60℃, and -70℃ corresponds to the saturation vapor pressure of water vapor 1.08 and 0.26Pa respectively; in both cases the driving force of the mass transfer has very little difference in regards to improve the drying rate. There is no practical significance.

(5) Refrigeration system: the task of the refrigeration system is to provide a low-temperature environment in the cold trap and/or in the drying chamber.

Here, the refrigeration system may be in various types: mechanical vapor compression refrigeration, absorption refrigeration, thermoelectric refrigeration, or using liquid nitrogen directly.

In the vapor compression refrigeration systems, the cold trap is the evaporator in the refrigeration system. For the cold trap with low temperature, the two-stage compression refrigeration cycle, or cascade refrigeration cycle may be adopted.

(6) The task of the vacuum system is to withdraw the "non-condensable" gases. In the temperature region dealing with freeze-drying, air, CO_2 and others are "non –condensable" gases. The non -condensable gases in the freeze drying system would come in two ways. One way is air leaking from the outside atmosphere into the vacuumed system; the other way is air and/or other gases escaping from the food or pharmaceutical material during drying process.

The specific volume of water vapor is very large, 8376 m^3/kg at -40℃, which means that sublimation of 1 gram of ice will produce near 8.4 m^3 of water vapor. So much water vapor is mainly condensed by the cold trap into liquid and eliminated from the system. There is still small amount of water vapor going into the vacuum system.

In addition to maintain vacuum degree and pumping speed, the vacuum pump in the freeze-drying system is also required to be applicable to water vapor, and a certain pumping capacity to eliminate water vapor. The requirement of vacuum degree is only about 1 Pa, which can be satisfied by rotary vane pump, or rotary vane pump with Roots pump in its low pressure side. The advantage of Roots pump is its larger intake pressure and greater pumping speed; its shortcoming is the relatively low pressure ratio. If the exhaust pressure of Roots pump is lower than atmospheric pressure, it is necessary to add rotary vane pump as a booster.

(7) Heating the materials inside the drying chamber is generally carried out through the plates as shown in Fig.1.4. Heating material through lower plate is by conduction; and heating materials through upper plate is by radiation. The electric heater or heat exchanger of a heat transfer agent can be installed inside the plates. The temperature of the plates can be controlled by a temperature control system.

1.5 Applications of freeze-drying technology

Looking at current technological development, the freeze-drying application may be approximately divided into following kinds:

(1) Freeze drying of foods and micro-organisms. Relatively speaking, the freeze drying technology of micro-organisms, coffee and milk is more mature; the freeze drying of fruits and vegetables is emerging and high value-added technology.

(2) Freeze-drying of biological pharmaceuticals. In the past 10 years, freeze-drying of biological drugs has become the most important application in the freeze-drying industry, and it is a field of most concern and with largest investment in freeze drying.

(3) Freeze-drying of human cells. If the human cells would be successfully freeze-dried, it will bring about a tremendous change in clinic medicine. The international academic community are very much concerned and actively studying the issue. However freeze drying of living cells is still at exploratory stage so far.

(4) Freeze-drying in other application domains.

1.5.1 Freeze-drying of micro-organisms

A large number of micro-organisms have been able to be successfully freeze-dried, such as bacteria, actinomycetes, yeast, filamentous fungi and viruses. The freeze-dried survival of micro-organisms is more than 80%. Nevertheless, up to now there are many micro-organisms can not be freeze-dried yet. Even for many micro-organisms which are freeze-dried successfully, the freeze drying procedures are not the same, and the protective agents are not so similar.

1.5.2 Freeze-drying of food

Freeze-dried food has marked advantages: keeping fresh color, smell and taste; avoiding the loss of nutritional ingredient and the surface hardening. Freeze-dried food is completely dehydrated, therefore it is light in weight, easy to rehydrate, fast soluble and can be stored at room temperature for long time. Freeze-drying is referred to as a new food processing technology of the 21st century.

Freeze-dried foods are high value-added products. In the current international market, the price of freeze-dried food is 7—8 times of the quick-frozen foods, and 4—6 times of the hot blast dry foods. Its economic value is quite obvious. Freeze-dried food is regarded as high-quality food, convenience food, leisure food. The technology is used in exported vegetables and fruits, or in travelling, exploration, navigation and so on[6,8].

However, freeze-drying still has its problems and difficulties. Freeze-drying process is a time-consuming and energy consuming process; the parameters of the freeze-drying process have decisive impact on the quality of freeze-dried food. Freeze-drying process, including the cooling, the sublimation drying, the desorption drying and the storage, is a very complex process of heat and mass transfer; and the process is closely related to the nature of the food. It is therefore necessary to determining the physical properties of foods, to understand the special heat and mass transfer processes and their effects on the quality of freeze dried foods. After that people can develop a reliable procedure for the optimization of freeze-drying process to produce high-quality food; and can make the freeze drying process a time-saving and energy-saving process, and reduce the costs thereof.

At present the proportion which freeze dried food occupies in the international market has achieved quite high level. Along with the enhancement of freezing and drying technology and the improvement of freeze-drying equipment, the price of freeze dried food is going down unceasingly.

1.5.3 Freeze-drying of pharmaceutical products

Most products of modern medicine are heat-sensitive, such as liposome, interferon, human growth hormone, as well as Chinese herbal medicine. These products are sensitive to temperature, mainly to higher temperature.

In the production of heat-sensitive drugs, to prevent the drug degeneration or the decreasing of drug quality by overheating, a widely used technology is freeze-drying. The pharmaceutical products made by this method have the following characteristics: structural stability, basically unchanged biological activity, almost no loss of the volatile ingredients and the heat sensitive ingredients. The pharmaceutical products have porous structure and good therapeutic effectiveness, high water loss of 95%—99%, convenient storage at room temperature or in the refrigerator for long time [2, 4, 9].

Specifically, when compared with other drying methods, the pharmaceutical products by freeze-drying technology have the following advantages:

(1) Drying drugs in low temperature will not have denaturation or biological vitality loss. The heat-sensitive drugs, such as liposomes, protein drugs, vaccines, bacteria, strain and blood products are particularly suitable to be preserved by freeze drying technology.

(2) The easily volatile constituent and easily heat denatured nutrition ingredient in the drugs have very few losses during freeze drying.

(3) The microorganism's growth and the enzyme function are almost impossible in freeze dried drugs.

(4) After freeze-drying the drugs can maintain their original volume and shape well. When rehydrated, their huge surface of porous structure will absorb water well and quickly restore and maintain the original shape.

(5) Drying is carried on at vacuum case; very little oxygen exists in the drying chamber, the easy oxidation constituents in the drugs can be protected easily.

(6) As 95% or more moisture contents in the drugs can be removed, the transportation and the long-term preservation of freeze-drying drugs become very convenient.

(7) Most freeze-dried drugs can be stored at room temperature; a few need be stored in refrigerator for extremely longer term preservation.

For most biological drugs, freeze-drying has become a very important procedure in their production. According to 1998 US statistics, 14% of the antibiotic class drugs, 92% of the macro-molecule biological drugs, 52% of the other biological drugs needed the freeze drying procedure in their production. In fact, freeze-drying procedure is employed in all the biological drugs developed in recent years. Moreover, as freeze-drying is the final stage in the manufacture of biological drugs, the quality of freeze drying process is playing a key role in the drugs quality.

In order to prevent the denaturation of protein and the damage of membrane structure of biological drugs, it is necessary to add appropriate protective agents for freeze-drying. The type of agents and the concentration are varies with different drugs. The type, concentration, and Ph value of the protective agent can have important influence on the quality of the freeze dried biological drugs.

1.5.4 Freeze-drying of human cells

If the human cells (such as red blood cells, platelets, cord blood cells and others) were able to be successfully freeze-dried, then people will have their own cells stored at home. Human cells are firstly freeze-dried into powders, sealed in glass bottles, and then stored at room temperature to preserve several years or decades. In case of emergency, the stored cells are able to revive by a simply rehydration.

Human cell freeze-drying attracts wide attention and is actively studied in the present international academic circle. If it can be successful, the technology will have extremely important applications and will bring about a significant change in the clinical medicine.

However, the freeze-drying of human cells is much more difficult than micro-organisms and biological drugs. Currently it is still at the exploratory stage and has not yet been put to clinical practice.

From the 1960s, some scientists have studied the freeze-drying to preserve the red blood cell. They experienced many failures and obtained preliminary success. To this day, the recovery of freeze-dried red blood cell is lower than 50%.

The study of freeze-drying of human platelets has continued about 40 years and a breakthrough was made not until the recent year. In 2001, Wolker W.F. and the others add trehalose into the formula, after rehydration of the freeze dried human platelets, the survival of platelets reached 85%. In 2003, they began to carry out a feasibility study of clinical application.

Human cord blood contains a large number of immature hematopoietic stem cells. Compared with the adult cell, the zero year old babies' immature hemopoiesis stem cells have the pollution-free, small allograft rejection, low immunogenicity, and its regeneration capacity and speed is about 10 to 20 times of the adults.

Since 2001, our exploratory study on freeze-drying experiment of human cord blood, whole blood and mononuclear cells (MNC) has achieved better results. The clinical purpose of freeze drying of the human cord blood is to preserve the $CD34^+$. The recovery of $CD34^+$ cells within the mononuclear cell population was assessed using fluorescein isothiocyanate conjugated $CD34^+$ antibodies. The recovery of $CD34^+$ cells reached 60%—68% [10—12].

1.5.5 Other applications of freeze-drying

In addition to the widespread application in food, medicine and organisms, freeze-drying technology has also applied in many new areas. These applications mainly employed the characteristics of freeze-drying, namely, material drying at low temperatures can maintain the original structure, the nature and the shape under their frozen state, which is unmatched against other drying methods [3].

Freeze-drying technology has found important applications in the preservation of skin, cornea, and bone, and in preparation of the scaffold of tissue engineering products. Freeze-drying technology is also applied to materials science for the preparation of various superfine powders, including ceramic powder, catalyst powder, metal powder, alloy powder and magnetic powder for recording. Freeze-drying technology has also found applications in the new health-care cosmetics, the biological sample preparation of electron microscopy, the sample manufacture of animal and human organs. Freeze-drying technology has demonstrated the characteristic in these application cases[4].

2

Fundamentals of Freeze Drying

2.1 Properties of water and aqueous solution

2.1.1 Phase diagram of water [13,14]

Pure water is a single component system. Its phase diagram is shown in Figure 2.1; and the characteristics of phase changes are shown in Table 2.1. There are three regions in the phase diagram, namely liquid state (water) , gaseous state (vapor) and solid state (ice), divided by three solid lines. Line OC represents the equilibrium states of water and vapor, namely the boiling or condensing curve; Line OB represents the equilibrium states of ice and water, the melting or freezing curve; Line OA represents the equilibrium states of ice and vapor, namely the sublimation or desublimaton curve. The point O represents the triple point of water in the temperature of 273.16 K (0.01℃) and pressure of 610.62 Pa (pascals), at which the three phases of water coexist in thermodynamic equilibrium. At normal atmosphere pressure (1.01325×10^5 Pa), the boiling temperature of pure water is 373.15 K (100℃) , and the freezing temperature is 273.15 K (0℃).

Table 2.1 The characteristics of water

Triple point	
temperature	273.16 K (0.01℃)
pressure	610.62 Pa
Critical point	
temperature	647 K (374 ℃)
pressure	2.2×10^7Pa
At normal atmosphere pressure (1.01325×10^5 Pa)	
boiling temperature	373.15 K (100 ℃)
melting temperature	273.15 K (0 ℃)

Point C (P_c=2.2×10^7 Pa, T_c=647K) is the critical point of water. A critical point, also called a critical state, specifies the conditions (temperature and pressure) under which a phase boundary between liquid and vapor no longer exists, the liquid density and the vapor density become equal, the heat of vaporization is zero, and there is no distinction between the two phases. For the states, where temperature is higher than the critical temperature T_c, water cannot be liquified with the increase in pressure. Usually, the gaseous states under the critical temperature are called "vapor"; and the gaseous states above the critical temperature are called "gas".

The extension line OD of line CD represents the vapor pressures of subcooled water at subzero temperatures, which is the metastable equilibrium line of subzero water and vapor. It is found in Fig. 2.1; the vapor pressure on the line OD is higher than that on the line OA at an identical temperature, therefore the subcooled water is unstable.

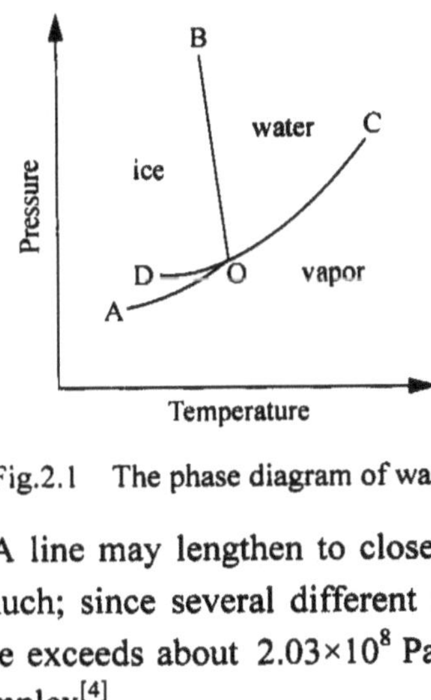

Fig.2.1 The phase diagram of water

Theoretically the OA line may lengthen to close to 0K; but the OB line cannot lengthen too much; since several different ice crystalline structures form when the pressure exceeds about 2.03×10^8 Pa, and the phase diagram becomes extremely complex[4].

Pressure increasing may reduce the freezing temperature of water, change the structure of ice crystalline, and even form the amorphous. However, the line OB in Fig.2.1 is very steep, the slop of the line OB in the vicinity of triple point O is about $dT/dP = -7.43\times10^{-8} K/Pa$. Very high pressure is needed to lower freezing temperature remarkably, and at the mean time the high pressure may cause many problems with respect to equipment, operation and safety.

In order to make possible amorphous drop of pure water, extremely rapid cooling rate of $10^6 K/s$ is needed, and the amorphous water drop is only in micrometer diameter.

2.1.2 Evaporation of water and sublimation of ice

For freeze-drying, the data of vapor pressure, vapor density and latent heat corresponding to equilibrium with liquid water and solid ice are all very important and useful. Table 2.2 and Table 2.3 list these data of vapor equilibrium with water and ice respectively.

Usually temperature range occurred commonly in freeze-drying is from about −100℃ to +40℃ or higher. Data are collected by 1K division in the temperature region from −45℃ to +40℃and 5K division in the region from −100℃ to −45℃. At the same time, in order to fully reflect the characteristics of water phase change, the temperature range of the data listed has been extended to the critical point

Table 2.2 The saturated vapor pressure, special volume of vapor and latent heat of evaporation equilibrium with liquid water[15]

temperature (℃)	pressure (kPa)	special volume of vapor (m^3/kg)	latent heat of evaporation (kJ/kg)
0.01	0.6106	206.356	2834.20
1	0.6571	192.456	2498.43
2	0.7060	179.770	2496.05
3	0.7580	168.027	2493.68
4	0.8135	157.138	2491.31
5	0.8725	147.033	2488.94
6	0.9353	137.645	2486.57
7	1.0020	128.948	2484.20
8	1.0728	120.851	2481.84
9	1.1481	113.327	2479.47
10	1.2280	106.329	2477.11
11	1.3127	99.813	2474.74
12	1.4026	93.744	2472.38
13	1.4978	88.089	2470.02
14	1.5987	82.816	2467.66
15	1.7055	77.898	2465.30
16	1.8184	73.308	2462.93
17	1.9380	69.022	2460.57
18	2.0643	65.018	2458.21
19	2.1978	61.273	2455.85
20	2.3388	57.773	2453.48
21	2.4877	54.500	2451.12
22	2.6448	51.434	2448.75
23	2.8104	48.563	2446.39
24	2.9851	45.873	2444.02
25	3.1692	43.351	2441.66

continued

temperature (℃)	pressure (kPa)	special volume of vapor (m^3/kg)	latent heat of evaporation (kJ/kg)
26	3.3631	40.986	2439.29
27	3.5673	38.767	2436.92
28	3.7822	36.683	2434.55
29	4.0083	34.727	2432.17
30	4.2460	32.889	2429.80
31	4.4959	31.161	2427.43
32	4.7585	29.536	2425.05
33	5.0343	28.007	2422.67
34	5.3239	26.568	2420.29
35	5.6278	25.213	2417.91
36	5.9466	23.936	2415.53
37	6.2810	22.734	2413.14
38	6.6315	21.600	2410.76
39	6.9987	20.530	2408.37
40	7.3835	19.521	2405.98
50	12.3499	12.029	2381.94
60	19.944	7.6697	2357.63
70	31.198	5.0402	2332.99
80	47.412	3.4055	2307.93
90	70.180	2.3592	2282.39
100	101.325	1.6718	2256.28
110	143.384	1.2093	2229.52
120	198.685	0.8913	2202.02
140	361.565	0.5085	2144.33
160	618.275	0.3069	2082.31
180	1002.871	0.1939	2014.87
200	1555.074	0.1272	1940.76
220	2320.1	0.08602	1856.2
240	3348.0	0.05964	1764.0
260	4694.0	0.04212	1660.2
280	6419.1	0.03010	1541.6
300	8591.7	0.02162	1403.0
320	11290	0.01544	1236.2
340	14608	0.01078	1025.5
360	18674	0.006970	722.6
374	22084	0.003482	111.5

Table 2.3 The saturated vapor pressure, specific volume and latent heat of sublimation equilibrium with ice[15]

temperature (℃)	pressure (kPa)	specific volume of vapor (m^3/kg)	latent heat of sublimation (kJ/kg)
-100	0.000001403	56947608.20	
-95	0.000003784	21729682.75	

continued

temperature (℃)	pressure (kPa)	specific volume of vapor (m^3/kg)	latent heat of sublimation (kJ/kg)
-90	0.000009672	8741258.74	
-85	0.00002353	3690036.90	
-80	0.00005473	1629195.18	
-75	0.0001220	749625.19	
-70	0.0002615	358551.45	
-65	0.0005406	177714.59	
-60	0.00108	90942.00	2836.27
-55	0.00209	48061.05	2837.13
-50	0.00394	26145.01	2837.80
-45	0.00721	14512.36	2838.29
-44	0.00811	13047.66	2838.37
-43	0.00911	11661.85	2838.44
-42	0.01022	10433.85	2838.50
-41	0.01147	9344.25	2838.55
-40	0.01285	8376.33	2838.60
-39	0.01438	7515.87	2838.64
-38	0.01608	6750.36	2838.67
-37	0.01796	6068.17	2838.70
-36	0.02004	5459.82	2838.71
-35	0.02235	4917.10	2838.73
-34	0.02490	4432.37	2838.73
-33	0.02771	3998.71	2838.72
-32	0.03082	3610.71	2838.71
-31	0.03424	3263.20	2838.69
-30	0.03802	2951.64	2838.66
-29	0.04217	2672.03	2838.63
-28	0.04673	2420.89	2838.59
-27	0.05174	2195.23	2838.53
-26	0.05725	1992.15	2838.48
-25	0.06329	1809.35	2838.41
-24	0.06991	1644.59	2838.34
-23	0.07716	1495.98	2838.26
-22	0.08510	1361.94	2838.17
-21	0.09378	1240.77	2838.07
-20	0.10326	1131.27	2837.97
-19	0.11362	1032.18	2837.86
-18	0.12492	942.47	2837.74
-17	0.13725	861.18	2837.61
-16	0.15068	787.49	2837.47
-15	0.16530	720.59	2837.33
-14	0.18122	659.86	2837.18
-13	0.19852	604.65	2837.02
-12	0.21732	554.45	2836.85
-11	0.23774	508.75	2836.68

continued

temperature (℃)	pressure (kPa)	specific volume of vapor (m^3/kg)	latent heat of sublimation (kJ/kg)
-10	0.25990	467.14	2836.49
-9	0.28393	429.21	2836.30
-8	0.30998	394.64	2836.10
-7	0.33819	363.07	2835.89
-6	0.36874	334.25	2835.68
-5	0.40176	307.91	2835.45
-4	0.43747	283.83	2835.22
-3	0.47606	261.79	2834.98
-2	0.51772	241.60	2834.72
-1	0.56267	223.11	2834.47
0.01	0.6106	206.356	2834.20

Based on these data, the vapor pressure equilibrium with liquid water and solid ice is drawn as (lg *P*-*T*) shown in Fig.2.2. And the vapor pressure and specific volume of vapor equilibrium with ice are drawn as (lg *P*-*T*-lg*v*) shown in Fig.2.3.

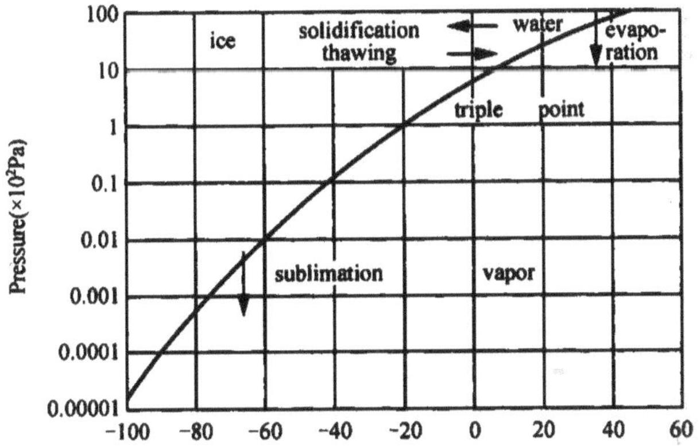

Fig.2.2 The vapor pressure equilibrium with water and ice (lg *P*-*T*)

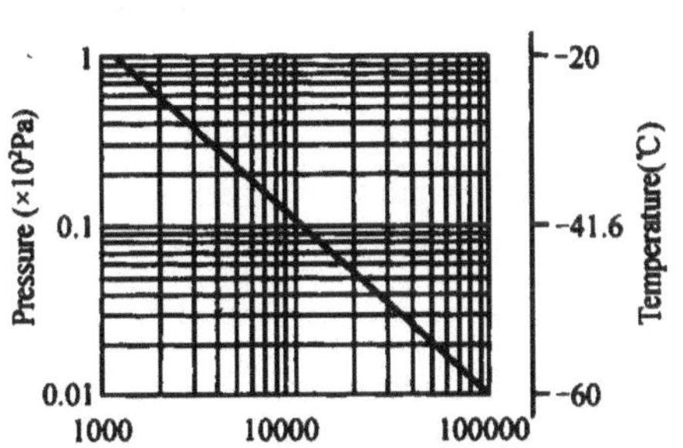

Fig.2.3 The vapor pressure and specific volume equilibrium with ice

In order to allow readers to have perceptual awareness of the latent heats of sublimation, evaporation, and melting, an enthalpy-temperature chart which includes ice, saturated water and saturated vapor, is shown in Fig. 2.4. It is clear from the chart that: both the evaporation heat and the sublimation heat are much larger than the melting heat. And at the triple point, the sublimation heat is the sum of the melting heat and the evaporation heat.

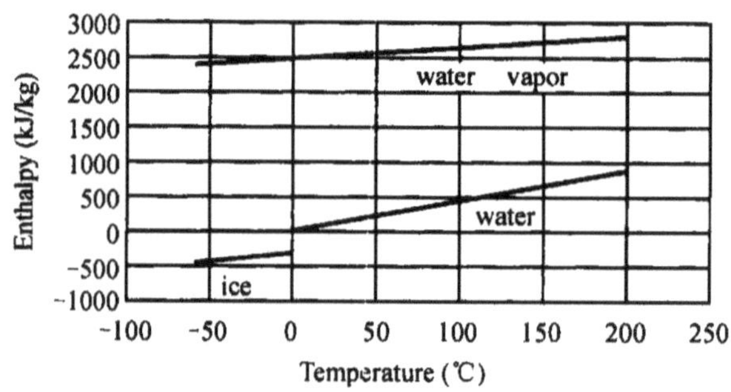

Fig.2.4 An enthalpy-temperature chart of ice, water and vapor (the enthalpy value of water at 0℃ is assumed zero)

2.1.3 Physical properties of aqueous solution

1. The composition and representation of aqueous solution

A solution is a homogeneous mixture composed of two or more substances. In such a mixture, a solute is dissolved in another substance, known as a solvent. Solution may be liquid and solid. All solutions are characterized by interactions between the solvent phase and solute molecules or ions that result in a net decrease in free energy. The ability of one compound to dissolve in another compound is called solubility. The physical properties of compounds such as melting point and boiling point change when other compounds are added. Together they are called colligative properties.

An aqueous solution is a solution in which the solvent is water. In many cases, such as in preparation of food and medicine, in freeze-drying process, and in cells and tissues in low temperature preservation, it is often necessary to prepare a variety of aqueous solutions with different solutes and different concentrations. There are several ways to quantify the concentration, such as mole fraction, molarity, molality, and parts per million (ppm). Therefore, we must firstly understand the several concentration expressions of solution, their precise meaning, as well as the conversion relations of these expressions [13].

1) Mole fraction x_s

The mole fraction of a solute in a solution is the relative proportion of mole number of solute to the total mole number of the solution. For each solute, the mole fraction x_s is the number of moles n_s divided by the total number of moles n_{total} in the solution, as shown in Eq. (2.1).

$$x_s = n_s/(n_w + \sum n_s) = n_s / n_{total} \tag{2.1}$$

here, n_s, n_w, $\sum n_s$, n_{total} is the mole number of a solute, water, total solutes, and the solution respectively. Mole fraction x_s is a dimensionless number.

As an example, if a mixture is obtained by dissolving 10 moles of NaCl in 90 moles of water, the mole fraction of NaCl in the aqueous solution is 0.1.

2) Molality m_s

Molality, a unit of concentration, is the mole number of the solute dissolved in one kilogram water (solvent), mol/kg_w. The value of molality is independent of temperature.

$$m_s = n_s / \text{kg}_w \tag{2.2}$$

The relations of mole fraction x_s and molality m_s are

$$\begin{aligned} x_s &= n_s/(n_w + \sum n_s) \\ &= m_s/(1/M_w + \sum m_s) \end{aligned}$$

and

$$x_s = m_s M_w /(1/+ M_w \sum m_s) \tag{2.3}$$

here, M_w is the molar mass of water, i.e., 18.015 g/mol = 18.015×10^{-3} kg/mol.

For very dilute aqueous solutions

$$x_s \approx m_s M_w \tag{2.4}$$

3) Molarity c_s

Molarity c_s is the molar number of solute dissolved in one liter of solution. The unit, therefore is the moles of solute per liter of solution, mol/L

$$c_s = n_s / V_{solution} \tag{2.5}$$

here, $V_{solution}$ is the volume of the solution, L.

The value of molarity is dependent of temperature. If the density $\rho\,(\text{kg/m}^3)$ of the solution at a state is known, the mass of the solution is ρV. The following equation (2-6) can be found.

$$c_s / x_s = \rho(n_w + \sum n_s)/(n_w M_w + \sum n_s M_s) \tag{2.6}$$

here, M_w and M_s (kg/mol) are the molar mass of the water and the solute respectively.

For dilute aqueous solution, there are

$$x_s = c_s \times M_w / \rho \tag{2.7}$$

and

$$m_s = c_s / \rho \tag{2.8}$$

In calculation, attention should be paid to the units, such as $M_w \approx 18 \times 10^{-3}$ kg/mol, and $\rho \approx 10^3 \text{kg/m}^3$ for very dilute aqueous solution。

4) Mass percent w_s

The mass percent w_s is defined as the ratio of mass of some solute to the total mass of the solution. There is

$$w_s = n_s M_s / (n_w M_w + \sum n_s M_s) \tag{2.9}$$

In the ways of representing concentrations, mole fraction and mass percent are dimensionless numbers, and molarity and molality yield dimensional quantities (per liter, per kilogram).

2. Colligative properties of dilute solution

Colligative properties are the properties of solutions that depend only on the solute quantity in the solution, or the concentration of solution, not on the nature of the solute. There are four colligative properties discussed here: vapor pressure decreasing, boiling point elevation, freezing point depression and osmotic pressure.

Measurements of these colligative properties for a dilute aqueous solution of a non-ionized solute such as urea or glucose can lead to accurate determinations of relative molecular masses. Alternatively, measurements for ionized solutes can lead to an estimation of the percentage of ionization taking place.

The dilute solution composed by the water and certain involatile nonelectrolyte solute is studied firstly.

1) Vapor pressure decreasing

The relationship between the lowering of vapor pressure and concentration is given by Raoult's law, which states that:

The water vapor pressure of the solution P_w is equal to the vapor pressure of pure water p_w^0 multiplied by the mole fraction of water x_w, or, the lowering of water vapor pressure of the solution $p_w^0 - p_w$ is equal to the water vapor pressure p_w^0 multiplied by the mole fraction of the solute x_s.

The relation is called Raoult's law as shown in Eq. (2-10).

$$p_w^0 - p_w = p_w^0 x_s \tag{2.10}$$

2) Boiling point elevation

At the same pressure, the boiling point of dilute solution T_b is higher than the boiling point of pure water T_b^0. The boiling point elevation is proportional to the molaliy of solution m_s, that is

$$\Delta T_b = T_b - T_b^0 = K_b m_s \tag{2.11}$$

$$K_b = RM_w\,(T_b^0)^2/r \tag{2.12}$$

here, gas constant $R = 8.314\text{J}/(\text{K}\times\text{mol})$; K_b is called as ebullioscopic constant; and M_w, r, T_b are the molar mass of water, the mole evaporation heat and the boiling point of pure water, respectively. $M_w = 18.0\text{g}/\text{mol}$; when T_b=373.15K, $r = 40.6\text{kJ}/\text{mol}$; it can be found $K_b = 0.51\text{K}\cdot\text{kg}/\text{mol}$.

3) Freezing point depression

When the temperature is lowing under constant pressure, if solid solution is not produced and the solid phase produced is from pure water, the freezing point of dilute solution is lower than the freezing point of pure water. The freezing point depression is proportional to the molality of the solution, that is,

$$\Delta T_f = T_f^0 - T_f = K_f m_s \tag{2.13}$$

$$K_f = RM_w\,(T_f^0)^2/L_f \tag{2.14}$$

where, K_f is called cryoscopic constant; L_f is the molar melting heat of pure ice at temperature of T_f^0. When T_f^0=273.15K, $L_f = 6.003\text{kJ}/\text{mol}$, and $K_f = 1.86\text{K}\cdot\text{kg}/\text{mol}$.

4) Osmotic pressure

Osmosis is a physical process in which a solvent (water) moves across a semi-permeable membrane by concentration difference .The semi-permeable membrane is permeable to the solvent (water), but not the solute. If a semi-permeable membrane separates a chamber of water and a chamber of solution, the osmosis is present; the water diffuses through the membrane from the water side to the solution side, which tends to reduce the difference in concentrations. This osmosis effect can be countered by increasing the pressure of the solution side, so that both sides have the same chemical potential. The osmotic pressure Π is defined to be the pressure required to maintain the equilibrium.

It can be proved that for the ideal dilute solution, the osmotic pressure Π is proportional to the mole fraction x_s of the solution. The relation is called Van't Hoff law as shown in Eq. (2.15).

$$\Pi = RTx_s / V_w^0 \tag{2.15}$$

here, gas constant $R = 8.314 \mathrm{J/(K \cdot mol)}$; T is absolute temperature, K; V_w^0 is the molar volume of pure water, $18 \times 10^{-3} \mathrm{L/mol}$ 。

3. Properties of actual aqueous solution

In actual aqueous solutions, these four phenomena: namely vapor pressure decreasing, boiling point elevation, freezing point depression and osmotic pressure, are all present. However they do not have the relations of colligative properties as in equations from (2.10) to (2.15).

As an example, the freezing point depression of actual aqueous solution is explained here. Regarding the dilute solution composed of non-electrolyte solute and water, the experiment has indicated its freezing point depression is proportional to the molality of the solution m_s, as shown in Eq. (2.13). If the solute is the electrolyte, it may dissociate into cation group and anion group, partially or completely. After dissociation, the three parts (the cation group, the anoin group and the undissociated molecules) may all make contribution to lower the freezing point. Therefore the value of m_s should be replaced by the sum of m_+, m_- and m_u. Here m_+, m_- and m_u are the molalities of the cation group, the anion group and the undissociated molecules, respectively.

When evaluate the impact of nonideality on vapor pressure of the actual aqueous solution, the activity a_w and activity coefficient γ_w are usually considered.

As "molality" is frequently used to indicate the freezing point depression of ideal solution, a similar concentration type called "osmolality" is used in non-ideal solution. An Eq.(2.16), similar to Eq. (2.13), is used for actual aqueous solution.

$$\Delta T_f = K_f \Omega \tag{2.16}$$

The value of K_f is the same as that in Eq. (3.13), which is $K_f = 1.86 \mathrm{K \cdot kg/osomolality}$. Ω osomolality/kg is the osmolality of the actual solution. And the ratio of the osmolality Ω and the molality m_s of the solution is called the osmotic coefficient.

2.1.4 Water activity and stability of the food and drug

1. Water activity in food and drug

The material in food and drug is approximately regarded as a solution. In the solution, water is the solvent, and the sugar, protein, carbohydrate and others are regarded as the solutes. If this solution at temperature T is in phase equilibrium with the air in the upper space, the chemistry potential of water vapor in the upper space should be equal to the water chemistry potential in the solution, according to the phase equilibrium condition. As shown in Fig.2.5, there is

$$\mu_w(\text{vapor}) = \mu_w(\text{food}) \tag{2.17}$$

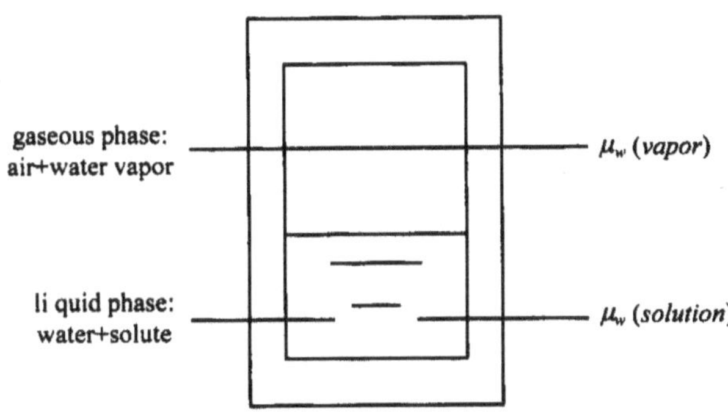

Fig.2.5 The vapor-liquid phase equilibrium of a solution in a adiabatic container

In the upper air, the chemical potential of water vapor $\mu_w(\text{vapor})$ is

$$\mu_w(\text{vapor}) = \mu_w^0 + RT\ln(f_w / P_0) \tag{2.18}$$

where μ_w^0 is the chemical potential of water vapor. When it is regarded as an ideal gas under a chosen standard pressure P_0 (for example, $P_0 = 101.325\text{Pa}$), f_w is the "effective pressure", or "fugacity", of the actual water vapor.

There is a relation between the effective pressure f_w and the actual water vapor pressure P_w.

$$f_w = r_w P_w \tag{2.19}$$

The coefficient r_w is called fugacity coefficient.

$$f_w \to P_w \text{ and } r_w \to 1, \text{ when } P \to 0 \tag{2.20}$$

In the solution of food and pharmaceutical material, the chemical potential of water vapor $\mu_w(\text{food})$ is

$$\mu_w(\text{food}) = \mu_w^* + RT\ln a_w \tag{2.21}$$

where μ_w^* is the chemical potential of pure water, and a_w is referred to the water activity in the solution.

Combing the Eqs. (2.18) and (2.21), there is

$$\mu_w^* - \mu_w^0 = RT(\ln a_w - \ln f_w) = RT(\ln a_w - \ln \frac{r_w P_w}{P_0}) \tag{2.22}$$

Since for pure water, $a_w = 1$, $r_w = 1$,

Therefore

$$\mu_w^* - \mu_w^0 = RT \ln \frac{P_0}{P_w^*} \tag{2.23}$$

Substitute the Eq. (2.23) into Eq. (2.22), the term P_0 can be eliminated, therefore

$$a_w = \left(\frac{r_w p_w}{p_w^*} \right)_T \tag{2.24}$$

2. Water activity and relative vapor pressure

When the air pressure is equal to the atmospheric pressure or lower, and the temperature is lower than 150℃, the values of r_w are between 0.9607 and 1.0[4]. It is the cases in freezing drying processes. Therefore

$$a_w \cong \left(\frac{p_w}{p_w^*} \right)_T = RVP \tag{2.25}$$

where p_w and p_w^* are the actual water vapor pressure in air and the saturated vapor pressure of pure water corresponding to the temperature T respectively. The ratio of p_w / p_w^* is the relative vapor pressure, RVP. Further, it is relative humidity of the air when the air is in thermal equilibrium with food or drug.

There is no device that can be put into a product to directly measure the water activity. However, the water activity of a product can be determined from the relative humidity of the air surrounding the sample when the air and the sample are at equilibrium. Therefore, the sample must be in an enclosed space where this equilibrium can take place. Once this occurs, the water activity of the sample and the relative humidity of the air are equal.

At present, most methods to measure the water activity of food are based on the measurement of relative humidity RVP of the air which is in thermal equilibrium with food. Several methods can be used to measure RVP, such as the air dew-point temperature measuring method, the psychrometer temperature measuring method, etc.

The concept of water activity comes from thermodynamics and physical chemistry, and already has been widely used. However, the value of water activity is still mainly determined by measuring the relative vapor pressure RVP. In recent years, many people suggest directly use RVP for the study of food and pharmaceuticals[16].

3. RVP and the stability of food and pharmaceuticals

Many reactions, such as microbial growth, enzymic reaction, nonenzymic reaction, may cause the deterioration, spoilage, browning and texture change of food. And several factors, such as RVP, the components, the temperature, the pH value, may affect these reactions. In which RVP is an extremely important factor, especially in unfrozen condition.

1) Influence of RVP on microbial growth

Rapid microbial growth is the most important contributor to the deterioration and spoilage of food, pharmaceuticals and cosmetics. For most microorganisms, RVP greater than 0.99 are most favorable to microbial growth. Most bacteria do not grow at RVP below 0.91, and most molds cease to grow at RVP below 0.80.

Certainly, the maximum permissible values of RVP for inhibiting reproduction of microorganisms may be different for different microorganisms' types and environment. But it is generally accepted that under the RVP< 0.6 condition, the microorganism cannot reproduce.

2) Influence of RVP on enzyme reaction

Enzyme reaction is another important contributor to the deterioration and spoilage of food. It is generally believed that when RVP=0.75—0.95, enzyme is most active. However the inhibition of emzyme action by lowering RVP vary from emzyme to emzyme. Decreasing RVP has obvious inhibition effect on the lipases, hydrogen peroxide enzymes; but it has no obvious effect on phosphatase. For amylase and a number of oxidase, the inhibition effect is most obvious when RVP = 0.65—0.70.

3) Influence of RVP on non-enzyme reaction

Foods containing proteins and carbohydrates are prone to non-enzymatic browning reactions, called Maillard reaction. The likelihood of Maillard reaction is not a monotonic function of RVP.

4) Influence of RVP on protein, vitamin and the texture of food

Some reports found that the milk powder with higher RVP losses large amount of lysine during storage. The protein change can be completely inhibited when RVP is controlled lower than 0.11; the loss of vitamin C can be inhibited when RVP <0.3; the loss of vitamin B1 is inhibited when RVP <0.2. RVP has also complex influence on the texture of food.

Food stability and RVP are closely related in many situations. The data in Fig.2.6 provide the examples of these relationships. Data in the figure reveal typical qualitative relationships between the reaction rate and RVP in the temperature range 25—45℃.

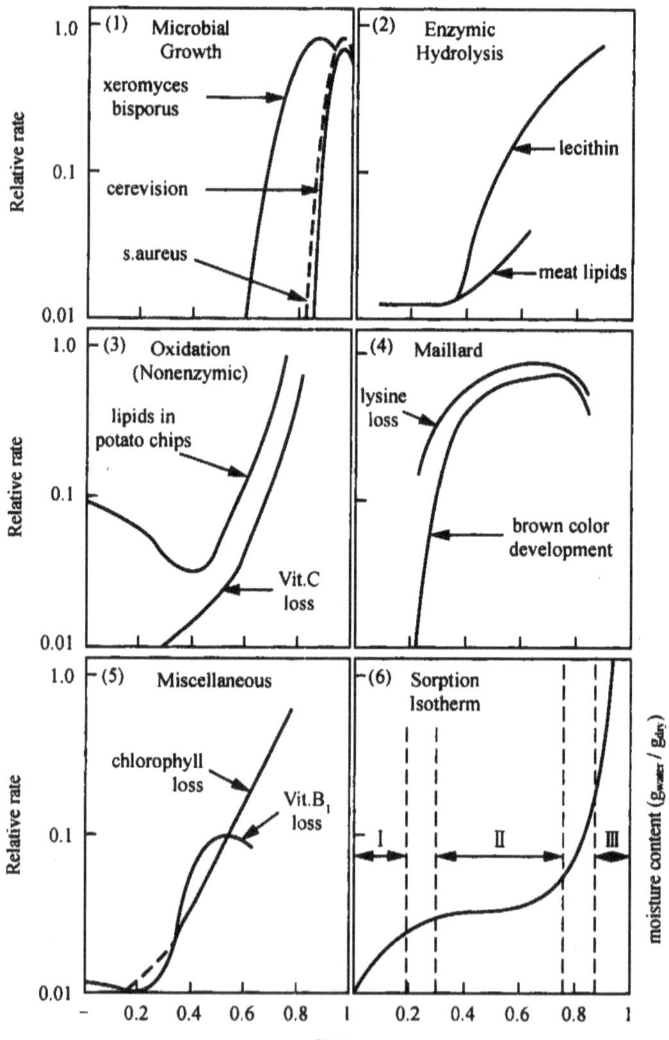

Fig.2.6 Relationships between reaction rates and RVP [16]

(1) Microbial growth versus RVP; (2) Enzymic hydrolysis versus RVP; (3) Oxidation (nonenzymic) versus RVP; (4) Maillard browing versus RVP; (5) Miscellanneous reaction versus RVP; (6) Water content versus RVP

Parts (2) to (5) in Fig.2.6 show several effects which cause the deterioration of food. The ordinates of these parts denote the relative rates of various reactions, all the abscissas denote RVP. The part (6) in Fig.2.6 shows water sorption isotherms, the ordinate here denotes moisture content g_{w}/g_{dry} .

Table 2.4 lists the maximum permissible values of RVP of several non-packaged dry food storaged in 20℃.

Table 2.4 The maximum permissible values of RVP of several non-packaged dry food storaged in 20℃ (Rao M.A., Rizvi S.S.H., Engineering properties of foods, New York: 2nd ed. Marcel Dekker,Inc.,1995)

Food material	RVP_{max}	Food material	RVP_{max}
sodium bicarbonate	0.45	milk powder	0.20—0.30
Crisp cakes biscuits	0.43	soup powder	0.60
whole egg powder	0.30	roasted coffee	0.10—0.30
gelatin	0.43—0.45	soluble coffee	0.45
Hard candy	0.25—0.30	starch	0.60
Chocolate cake	0.73	wheat products	0.60
milk chocolate	0.68	suger	0.55—0.92
potato chips	0.11	dehydrated meat	0.72
flour	0.65	dry fruit	0.60—0.70
oat powder	0.12—0.25	dewatering pea	0.25—0.45
beef juice powder	0.35	air-dried bean	0.08—0.12
skim milk powder	0.30	Orange powder	0.10

According to the above analysis, it is concluded that RVP (or moisture content activity) has vital significance to the stability of food and pharmaceutical. In general, low RVP is extremely advantageous in maintaining the stability. However attention should still be paid to the following situations:

(1) Even if in a very low RVP situation, some other factors, such as fat oxidation, still exist and cause the deterioration of food. In order to maintain stability for a long time, some other technologies must also be applied, such as low temperature and vacuum.

(2) The maximum permissible values of RVP listed in Table 2.4 are obtained in certain conditions, including temperature, PH value, etc. Moreover these values mostly only serve certain target, like some kind of microorganisms. If the conditions or the targets change, these data will no longer be applicable.

2.1.5 Vitrification of aqueous solution

Since the 1980s, scientists have begun research on vitrification problem of aqueous solution in the fields of cell cryopreservation, food freezing and freeze drying[6,13,16—20]. In freeze-drying, the vitrification issue not only relates to the solution formula and the determination of cooling temperature, but also has extremely important relations with transit temperature control, collapse prevention, the quality and stability of the freezing dried product. Solution vitrification has become a key issue and a research hot spot of the modern freeze-drying technology of food and pharmaceutical. Many various issues of vitrification will be discussed in later chapters of this book. The most basic concepts and the nature of the aqueous solution vitrification will be introduced firstly in this section.

1. Non-crystalline state and vitrification process

In nature, the solid state of substance has two forms: crystalline state and non-crystalline state (amorphous state). In macroscopic view, both states present solid character, which maintains a fixed volume and shape; whereas the distinction lies in its internal the microscopic structure. In the non-crystalline state, the microscopic particle arrangement is irregular, thus it is also called amorphous state (amorphous).

There are three main categories of non-crystalline materials: metal, inorganic and organic non-crystalline materials. As people get used to call the silicate non-crystalline solid "glass", the term "glassy state" is gradually expanded to represent all the non-crystalline solid states.

Solidification of cooling liquid can be achieved in two different ways: One way is that the liquid passes through a phase change, and it is solidified discontinuously to crystalline state. This process is called the crystalization phase change process. The other way of solidification is that the liquid does not go through a phase change, instead it is solidified continuously to non-crystalline (glassy) state. This process is called vitrification or glass transition.

Crystallization process is a first order phase change process: the crystallization process happens at an exact temperature (is called crystallization temperature). In the process, the latent heat of phase change is released; the discontinuous changes of volume and entropy, before and after the phase-change, have taken place. Since volume and entropy are the first derivatives of Gibbs free energy, the crystallization phase change is also called "the first order phase change".

$$V = \left[\frac{\partial G}{\partial P}\right]_T ; \quad S = -\left[\frac{\partial G}{\partial T}\right]_p \tag{2.26}$$

However in the vitrification process, there is no latent heat released, no non-continuous changes of the material properties; all the changes are continuous. The only change during the process is the change of state, i.e., from liquid state to solid state (non-crystalline solid state). In conclusion, the vitrification process is a state transition process.

From the perspective of thermodynamics, crystalline state is stable and non-crystalline state is metastable. Nevertheless the non-crystalline state has enormous viscosity; the molecular activity inside the glassy solid is nearly zero; the vitrification solid state is an amorphous state bereft of fluidity. Therefore within a limited, practical time scope, vitrification solid state can not turn into a crystalline solid and it is very stable in practice.

Fig. 2.7 gives the relation of dynamic viscosity of some polymers with the temperature. When temperature drops, viscosity rises dramatically. The ordinate denotes the denary logarithm of the dynamic viscosity; the abscissa denotes the temperature ratio of T_m / T, where T_m is the melting temperature of the polymer. The SI physical unit of dynamic viscosity is pascal-second (Pa·s), which is identical to $kg/(m \cdot s)$ [17].

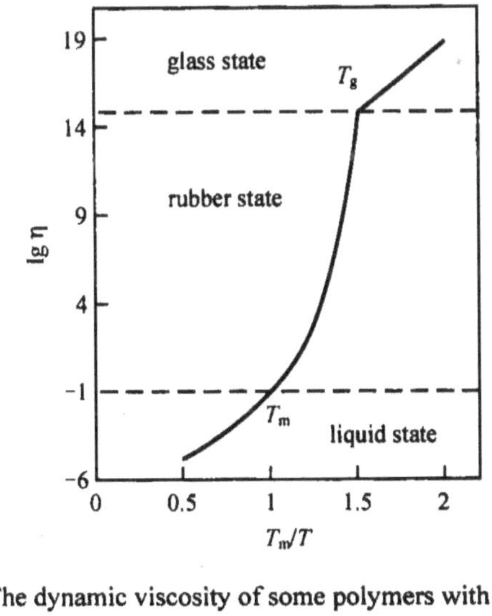

Fig.2.7 The dynamic viscosity of some polymers with temperature

Now the majority of scientists have agreed to take viscosity $\eta \geqslant 10^{14}\,\mathrm{Pa\cdot s}$ as the judgment of glassy state condition, and the temperature which corresponds $\eta = 10^{14}\,\mathrm{Pa\cdot s}$ of some material as the vitrification transition temperature (glass transition temperature) T_g. The glassy state corresponds to the area of $T < T_g$.

2. Vitrification and state diagram of aqueous solution

1) Vitrification of water and aqueous solution

Since vitrification transition is a non-equilibrium dynamic process, the formation of glassy state is mainly decided by dynamics factor, such as cooling rate. As long as the cooling rate is quick enough, and the final temperature of cooling is low enough, the majority of materials can be transited from liquid state to glassy solid state. Here, "quick enough" cooling means that there is not enough time for crystallization to take place when substance passes through the crystallization area; the "temperature low enough" means that the final temperature is lower than the vitrification transit temperature T_g.

In different systems, the necessary cooling rates for vitrification are quite divergent. For SiO_2, a cooling rate of $10^{-3}\,\mathrm{K/s}$ is enough for the vitrification of a plate with thickness of 200cm; for GeO_2, only 0.35 K / s cooling rate is needed to vitrify a plate with thickness of 3.5cm.

The viscosity of pure water at room temperature is as low as about $1.00\times10^{-3}\,\mathrm{Pa\cdot s}$ (20℃); and the glass transition temperature of water is about −135 ℃, that is $T_g = -135$ ℃, which corresponds to $\eta = 10^{14}\,\mathrm{Pa\cdot s}$. It is very difficult to vitrify pure water and even when the cooling rate reaches as high as $10^7\,\mathrm{K/s}$,it is also very hard to achieve glassy solid of 1 μm in thickness.

In the fields of biology, medicine, food and pharmaceuticals, the concept of vitrification was firstly applied to low temperature preservation of biological cell and tissues, such as blood cells, embryos and so on. Afterwards the vitrification theory has obtained widespread applications in the food science, mainly in low temperature vitrification preservation of foods. In recent years the vitrification transition theory has begun to apply to the stability studies of the pharmaceuticals freeze-drying, and it is discovered that the final temperature of cooling should be lower than its vitrification transition temperature. In the drying processes, the vitrification transition temperature elevates unceasingly when water content reduces. In all the primary drying, the secondary drying, and in the storage, the temperatures of pharmaceuticals in these processes should be lower than their corresponding vitrification transition temperatures.

2) State diagram of aqueous solution

A kind of state diagrams for aqueous solution is developed for convenience's sake. These diagrams are not called phase diagrams, because the curves in the diagram not only cover the lines that reflect phase changes, such as the freeze - melting curve T_m, the solution precipitation curve, but also cover several state transition lines, such as glass transition curve T_g and devitrification curve T_d. Such state diagrams are sometimes referred to as "supplementary phase diagrams " or "solid-liquid state diagrams".

Fig. 2.8 is simplified state diagram of sucrose-water solution; it only reflects the equilibirum phase changes of melting and the non-equilibrium state transition of vitrification.

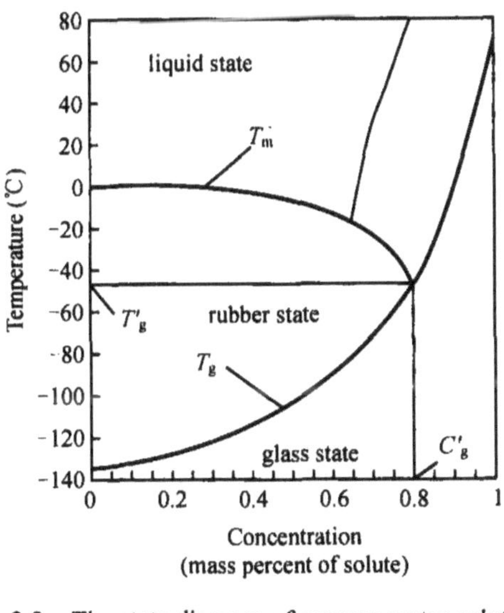

Fig.2.8 The state diagram of sucrose-water solution

As can be observed from Fig.2.8, the melting temperature of pure water is $T_m = 0$ ℃ and the melting temperature T_m of an aqueous solution is decreasing with the rise of concentration. The glass transition temperature of pure water is $T_g = -135$ ℃ and the glass transition temperature of an aqueous solution T_g is increasing with the rise of concentration. At the intersection of these two curves, the corresponding concentration C_g' and the corresponding temperature T_g' are known as the glass transition concentration and the glass transition temperature of the maximally frozen-concentrated state respectively.

The values of T_g' and C_g' of an aqueous solution are dependent on the nature of the solute, especially on the glass transition temperature of the solute T_{gs}. The value of different solutes are quite divergent, for example, $T_{gs} = -93$ ℃ for glycerin; $T_{gs} \approx 70$ ℃ for sucrose, and $T_{gs} = 101$ ℃ for lactose.

3. The ways to realize vitrification of aqueous solution

The liquid solution possibly achieves two kinds of glass condition: One is the complete glass condition; the other is the glass condition with partial crystallization, or the partial glass condition. Through a very rapid cooling, the thin particles of dilute solution can be vitrified. The required cooling rates are the orders of magnitude as high as $10^4 - 10^6 \mathrm{K/s}$.

Because of reasons like high heat capacity of water, complete glass condition is merely possible for very small samples. In practice, complete vitrification of liquid dilute solutions of the food and the drugs is almost impossible, only the partial glassy condition can be achieved.

At present, there are 3 ways for aqueous solution to achieve glassy state. It is illustrated in Fig. 2.9.

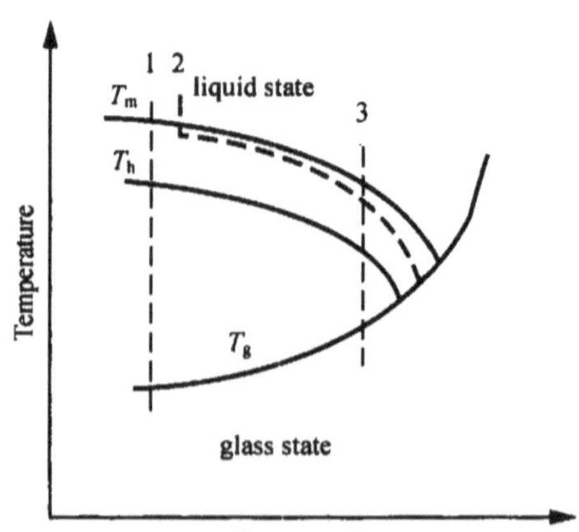

Fig.2.9 The 3 ways to achieve the glass state

(T_m -the melting temperature; T_h -the homogenous nucleation temperature; T_g -the glass transition temperature)

Way 1: for dilute solutions, T_g must be much lower than T_m, the temperature difference may reach 20—100K or more. Only extremely high cooling rate can cause the biological sample to be coo led to temperatures below T_g, and the cooling rate is so fast that the ice crystal has not enough time to form during the cooling period. The complete vtrification can be achieved by this way.

For aqueous solutions with concentration less than 20%, the critical cooling rate, which the complete vitrification needs, is generally in the order of 10^2-10^3K/s. This method is called "the complete vitrification method", or "the extremely quick cooling method", it is suitable for very small samples only (for example, the diameter of the sample is smaller than 1mm).

Way 2: cooling dilute solution in "two steps". The first step of cooling is carried out at ordinary rate; ice crystals form at the outside of the cells, or the outside of the matrix of the food or pharmaceuticals; water transports through the membrane of cells (or the membranes of the matrix) from the inside of cell (or matrix) to the outside. The concentration inside the cell (or matrix) is gradually increasing, therefore the glass transition temperature of the inside solution is increasing and the temperature difference (T_m-T_g) is decreasing too. The second step is to decrease the temperature at high rate to realize solution vitrification inside the cell (or matrix). This approach is also known as "the partial vitrification method".

Way 3: If the concentration in food and pharmaceuticals is as high as close to the value of C_g'; and the temperature difference (T_m-T_g) is small, then the vitrification can be achieved ordinary cooling rate (for example, the sample is plumped into the liquid nitrogen directly). This approach of vitrification is called "high concentration method". In cryobiology, this kind of high concentration solution is known as "vitrification solution".

In general, a solution with certain initial concentration starts cooling from the room temperature; formation of ice crystal begins at some temperature between the melting temperature T_m and the homogeneous nucleation temperature T_H. The released latent heat of crystallization forces the sample to return to the melting temperature T_m. Then the solution splits out ice crystal unceasingly along the melting temperature line; and the concentration inside the unfrozen solution is increasing continuously. Till arriving at the intersection of the melting curve and the vitrification curve, the solution reaches maximal frozen-concentrated state; the concentration of the solution is enhanced to C_g'. When cooling goes on, then the water inside the unfrozen solution will no longer be crystallized but become the amorphous state. The final solid state is a mosaic construction of glassy matrix and ice crystals.

4. Relation of stability and vitrification

In recent decades, along with rapid development of frozen food, cryopreservation, freeze-drying and vitrification theory especially, scientists proposed some stability theories regarding food and pharmaceuticals, such as vitrification theory and food polymer theory. These theories mainly came from research on material stability in the glass state.

Food polymer science was extended from the theory of engineering materials science. In the 1950s, scientists put forward a synthetic polymer science to study the engineering polymeric materials, including the glassy phenomenon and the glass transformation. They proposed that the cooled polymeric materials can be divided into three areas namely liquid area, rubber-state area, and glass-state area, as shown in Fig.2.7. In general, the bigger the polymer relative average molecular weight of the polymer, the higher its viscosity, the easier to reach the glassy state.

Since the 1980s, scientists have explored the possibility of synthetic polymer science application in other fields, such as biological materials, living tissues, food and pharmaceuticals.

The majority of biological tissues, foods and pharmaceuticals can be taken as "the biopolymer"; and they include often the amorphous glass state, especially in the conditions of dried, frozen and freeze drying.

In cold storage of a high quality food, it is the best to maintain the temperature of the refrigerator lower than the glass transition temperature T_g of the food. For freeze drying of pharmaceuticals, all temperatures of the material during the freeze-drying process and in the storage period should be below the glass transition temperatures corresponding to varied concentrations.

5. Indication of freeze-drying processes in the state diagram

Under ultra-rapid cooling, a small sample is expected to cool down directly to the area below the glass transition temperature T_g, and realizes the complete vitrification. The cooling process carries on along the curve g , like the line segment ABCK as shown in Fig. 2.10[16].

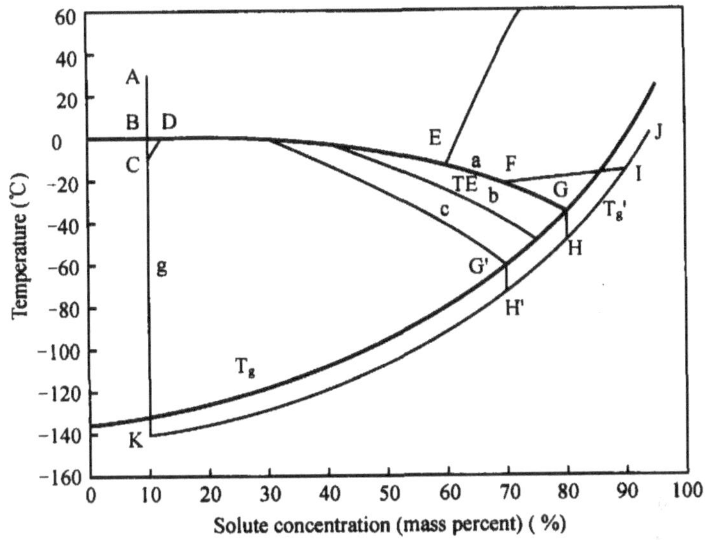

Fig. 2.10 Indication of freeze-drying process in a state diagram

Under overwhelming majority of situations, limited rate of cooling cannot bring about the sample to realize complete vitrification. The actual cooling process carries on along one of the curves a, b, c and others. Obviously, the order of the cooling rates is: g > c > b > a. In a process with higher cooling rate, there are more glassy state proportion and less crystal ice proportion in the final solid state.

Now we are to discuss firstly the freeze drying process with a slow cooling rate. A sample is cooled from room temperature at point A, crossing freezing point B, and subcooled to point C. At point C, the nuclei of ice are formed; and the sample temperature returns to freezing temperature D. As the temperature drops, the ice forms incessantly, and the concentration inside the unfrozen sample increases continuously. Theoretically, when arrives at the eutectic point E (T_e), all the constituents of the sample crystallize simultaneously. However, for the majority of biological products, very few solutes will crystallize at the eutectic temperature. Hence if the temperature continues to decrease, more ice crystals will form; and the solute concentration inside the unfrozen solution will be enhanced to be supersaturated, along the line EFG, till T_g', then the glass transition takes place. At this point the supersaturated liquid solution is transited to glassy state, and the whole sample becomes a complex mosaic solid structure, in which there are ice crystals and glassy state solution. Further cooling from the point of G to the point H, the temperature drops to temperatures below T_g', however the concentration is invariable.

An ideal drying process should be carried out along the line HIJ, that is, in the process of drying the sample temperature should be always maintained below the corresponding glass transition temperature T_g. In these cases, the molecular fluidity is very weak and the material is highly stable, and the collapse and other undesirable phenomena can be avoided. When there is insufficient cooling, for example, the sample is cooled to point F (which is higher than the glass transition temperature at that concentration), and the heating starts for drying, the drying process will be carried out along the line segment FIJ. At the initial stages of drying, the collapse will not appear owing to the existence of ice crystal; but as long as some of the material has dried up, the ice has disappeared, and the temperature is above the glass transition temperature, the collapse may take place as the heating continues.

When eutectic is formed in cooling process, the heating temperature in the sublimation drying process (primary drying process) should be kept merely lower than the eutectic temperature T_e. For the majority of materials, the heating temperature of sublimation drying should be lower than the glass transition temperature T_g. The valve of T_g increases along with the concentration.

In the attempt of elevate the transition temperature T_g, scientists are diligently seeking high polymer chemical additive. In this way, the heating temperature can be raised, and the time and energy can be saved in the whole process of drying.

2.2 Freezing process

2.2.1 Typical temperature decreasing curve

When a sample (a plastic bag or an ampoule of pure water) is put in the bath of a programmable cooler; and there is good thermal contact between the bath and the sample. The sample temperature is shown as the solid line in the Fig.2.11 as the bath temperature decreases from point “A” at a constant cooling rate as shown as the dotted line in the figure.

The normal freezing point "B" of pure water corresponds to 273.15K. However in most cases the freezing of pure water is not generally initiated at the equilibrium freezing temperature 0℃, but at some temperatures (for example point "C") lower than 0℃. This kind of phenomenon is called supercooling. The temperature difference between the phase equilibrium freezing temperature and the temperature of ice nucleation is called the degree of supercooling, or simply "supercooling".

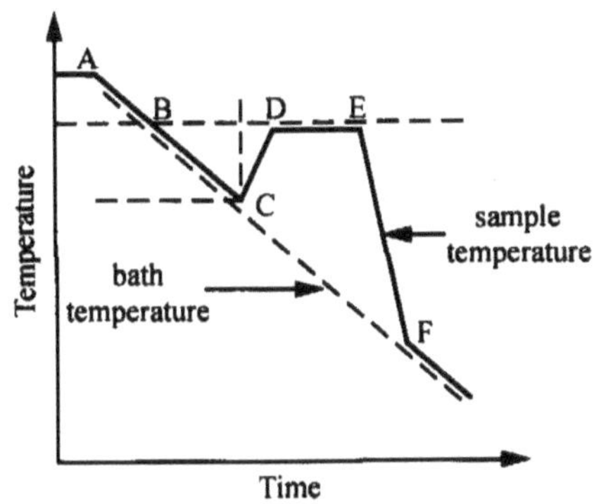

Fig. 2.11 The temperature decreasing curve of pure water

In the cooling path ABC, the sensible heat of the sample is released and the sample temperature tracks the bath temperature closely. At point C, the ice nuclei are formed and the ice crystals are growing. The release of freezing latent heat tends to elevate the sample temperature to its equilibrium freezing temperature, which is path CD. Along path DE, ice crystals are formed and latent heat is released unceasingly; the sample temperature remains constant at freezing temperature. After water is completely frozen at point E, the sample temperature begins to decrease as sensible heat is extracted from it, which is tracked by path EF. The cooling rate during the path EF may be substantially higher than the cooling rate of the bath, which is an issue of serious concern in the cryopreservation of biological cells and tissues. To control the cooling rate in path EF is an important technology of thermal control in cryopreservation[13,22].

2.2.2 Supercooling and homogeneous nucleation

In fact there are two mechanisms in the process of water freezing: ice nucleation and ice crystal growth. The ice nucleation process is mainly decided by thermodynamics condition; and the ice crystal growth process by the dynamical condition.

The states between B and C in Fig.2.11 are metastable in which the nuclei may be initiated by one of the two modes of nucleation: homogeneous nucleation or heterogeneous nucleation.

Homogenous nucleation refers to situation where the probability of nucleation is equal anywhere in a system. Thermal fluctuation possibly causes atoms or molecules gathering to form the clusters, which are embryos or nuclei of new phase (ice crystal). The clusters need to reach a critical size in order to become stable nuclei. Such critical size is dictated by the operating conditions (supercooling, etc.).

Fig.2.12 expresses the relation of the formation energy of solid phase with the radius of solid particle. The critical dimension of the solid phase can be denoted by its radius r^*. When $r > r^*$, the clusters are called nuclei. As nuclei grow, the nucleation proceeds, and the freezing becomes a spontaneous process. When $r < r^*$, the clusters are called embryos. Nucleation will not proceed, and the ice cluster will voluntarily vanish.

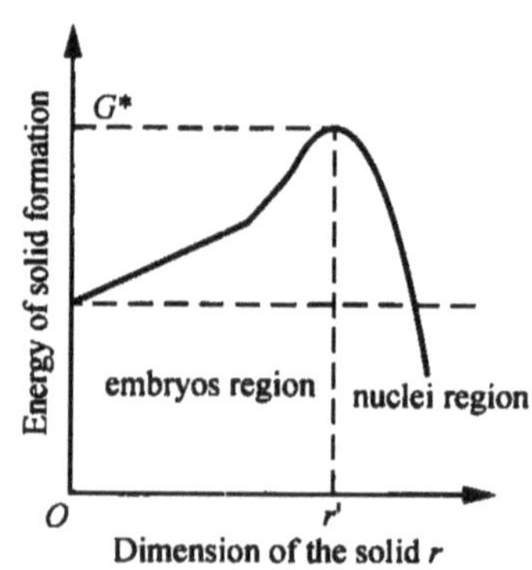

Fig.2.12 The unstable embryos and the stable nuclei in ice nucleation

When supercooling of water is enhanced, the deviation from the equilibrium state is enlarged. The critical radius r^* and the critical energy of nucleus formation G^* drop dramatically with the supercooling, as shown in Fig.2.13. The probability of nucleation increases exponentially with supercooling. For a water droplet of 1 μm diameter, homogenous nucleation occurs at approximately −40 ℃.

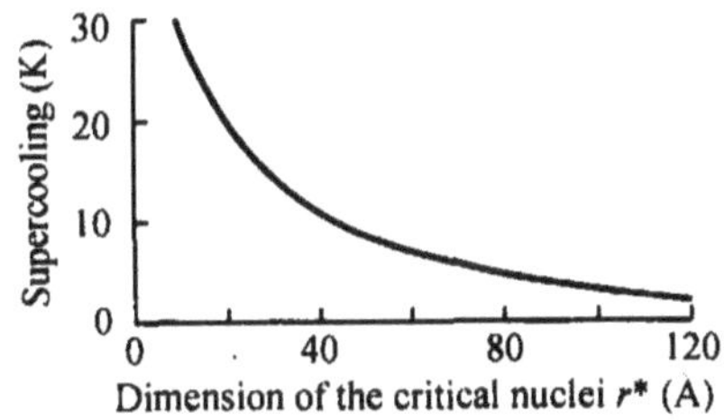

Fig.2.13 The critical dimension r^* of nuclei with the supercooling [13,23]

2.2.3 Heterogeneous nucleation

Homogeneous nucleation take place in pure systems consisting solely of pure water. In samples of water which are not ultra pure, heterogeneous nucleation takes place in preference to homogeneous nucleation. Heterogeneous nucleation occurs when water molecules aggregate in a crystalline arrangement on nucleants (catalysts), such as suspended foreign bodies, dust particles, surface film, the wall of container. Since bulk water is rarely supercooled to more than a few degrees below its melting point, homogenous nucleation is presumably not the common mode of nucleation, and the heterogeneous nucleation is of greater importance.

2.2.4 Crystal growth

The crystallization process consists of two major events, nucleation and crystal growth. Crystal growth is the subsequent growth of the nuclei that succeed in achieving the critical cluster size. Nucleation and crystal growth concur where there is supercooling. Supercooling is the driving force of crystallization. Depending upon the conditions, either nucleation or growth may be predominant over the other, and as a result, crystals with different sizes and shapes are formed.

The nucleation rate J [$\mathrm{new \cdot nuclei/(m^3 \cdot s)}$] refers to the new nuclei produced in the unit time and unit volume under certain temperature; the crystal growth rate U [$\mathrm{mm/min}$] refers to the linear growth rate of the crystalline under certain temperature. Fig. 2.14 sets forth an example of 50% (volume fraction) PVP aqueous solution. The nucleation rate J in the figure is in two dimensions, that is [$\mathrm{new \cdot nucei/(m^2 \cdot s)}$]. Some phenomena are illustrated in the figure.

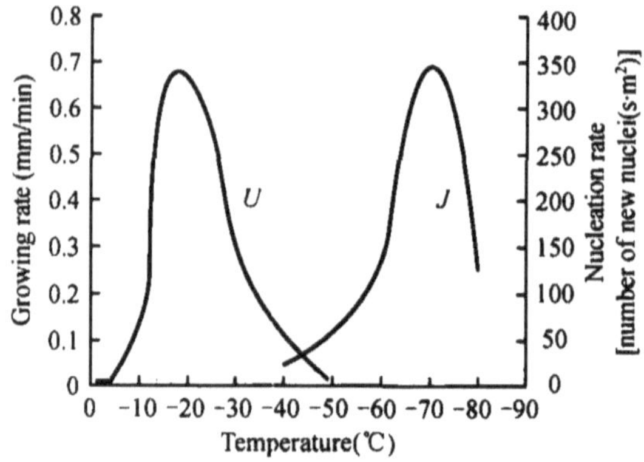

Fig.2.14 The nucleation rate J and crystal growth rate U of 50% PVP aqueous solution with temperature[24]

(1) The freezing temperature of 50% PVP aqueous solution $T_m \approx -5$ ℃, the nucleation and the crystal growth both appear in the supercooling area. And their maximum peaks correspond to different temperatures.

(2) The two events of crystallization process, nucleation and crystal growth, occur simultaneously, but their rates are quite different at varied temperatures. If the crystallization process occurs mainly at the region where the nucleation rate J is higher and the crystal growth U is lower, a large number of small and fine crystals are formed. However when the crystallization process occurs mainly at the region where the nucleation rate J is lower and the crystal growth rate is higher, a few large and coarse crystals are formed.

(3) The two curves J and U usually have an overlap area which is called "the crystallization area". That area is most advantageous to crystallization where the values of J and U are both high. If the crystallization area of an aqueous solution is bigger, the aqueous solution is easier to crystallize; otherwise the aqueous solution is easier to vitrify.

From the viewpoint of dynamics, whether the final state of cooling of a solution is crystalline or in vitreous state depends on the competition of two rates: One is the rate of nucleation and crystal growth; the other is the rate of temperature drop of the sample. The latter is decided by the sample quantity, its geometry shape, thermal conductivity, heat capacity, as well as the outside cooling condition; the former is decided by the thermal driving force of phase change, which is the Gibbs free energy difference, and the molecular mobility; they are all related to the degree of supercooling.

In order to study the freezing phenomenon thoroughly, some kinds of cryomicroscope system have developed, and several freezing phenomena under different cooling rates of pure water, and the aqueous solutions containing DMSO, sucrose, sodium chloride, have been studied. Some new rules of freezing aqueous solutions have been found, such as the air bubble formation and its distribution, the influences of the cooling rate and the cryoprotective agents on the crystalline structure, and the influence of stronger electric field on the freezing[25—32].

For cryopreservation of cells and tissues, different cells need different cooling rates. For freeze drying of pharmaceuticals and foods, various cooling rates are also needed. In order to meet these requirements, several cooling technologies have been developed, such as two–step cooling by commercial low temperature refrigerators and manual cooling using a liquid nitrogen container. These cooling methods are simple and ease of operation, however they cannot complete complex cooling programs. The programmable coolers (or called programmable freezer), developed along with the computer technology, have already become the essential equipment to realize cryopreservation of biological materials. According to the principles of temperature drop and controlling, the programmable coolers may divide into two kinds: jet type of liquid nitrogen (or called spurt type) and immersion type of liquid nitrogen. The immersion type cooler is using the temperature gradient inside the neck of the liquid nitrogen container; and the sample is moving inside the neck by step motor. Reader who are interested in relevant information may look at the following references [4, 34—39].

In order to realize vitrification of dilute solution, very high cooling rate is needed. The critical cooling rate for vitrification may be estimated based on the constituents of aqueous solution.[40, 41] Quenching of small samples into liquid nitrogen is a conventional method reaching high cooling rates. The cooling rate achieved closely relies on the mass of sample, the speed and the depth of quenching, as well as the supercooling of liquid nitrogen[42—46].

To realize super cooling rate, small sample can be quenched fast into liquid nitrogen. The rate of cooling is not only relevant to the sample's material volume and quality but also in close relationship with the supercooling degree of liquid nitrogen, the speed of quenching and the depth of quenching.

2.3 Sublimation drying process

2.3.1 Heat and mass transfer process during the drying step

In the analyses of freeze-drying process, there are two kinds of moisture content of material to be considered. One is freezable at low temperatures, which may be called "the free water" or "the capture physically water"; the other is unfreezable at low temperatures, which is also called "the bound water". Regarding the materials of high moisture content, "free water" is generally above 90% of the total moisture content.

Sublimation drying, also called primary drying, is heating frozen material inside the drying chamber; and the chamber is vacuumed by a vacuum pump. During this step, the frozen water inside the material is sublimated directly to water vapor; and the material is drying. The sublimation starts from the outside surface and passes inward gradually; the porous dried layer is formed in the same way simultaneously. The interface of frozen layer and the dried layer is called the sublimation front or the ice front. The heat required by the sublimation may be supplied in several ways: heat radiation through porous dried layer, heat conduction through frozen layer, or microwave heating the material directly. Vapor escapes from the ice front through the porous dried layer.

Therefore, heat transfer and mass transfer occur at the same time in sublimation step. When ice crystals in material are completely sublimated, the first step finishes. At this stage, above 90% of the initial moisture content has been removed.

2.3.2 Several typical modes of heat transfer and mass transfer in the drying process

Fig.2.15 shows common modes of heat transfer and mass transfer in the drying process.

(a) the heat rate Q required is acquired by conduction from the base of the frozen layer; the vapor escaping from the sublimation front at rate G is from the top of the dried layer by diffusion.

(b) the heat rate Q is acquired from the top of the porous dried layer; the vapor rate G is also from the top of the dried layer in the opposite direction of Q.

(c) the heat rate Q is acquired from the top of the porous dried layer and the base of the frozen layer; the vapor rate G is from the top of the dried layer. The mode occurs often in the freeze dryer with multi-layered shelves; the material on a shelf absorbs the conduction heat from the shelf beneath, and the radiation heat from the above shelf. In most cases, the former is much larger than the latter.

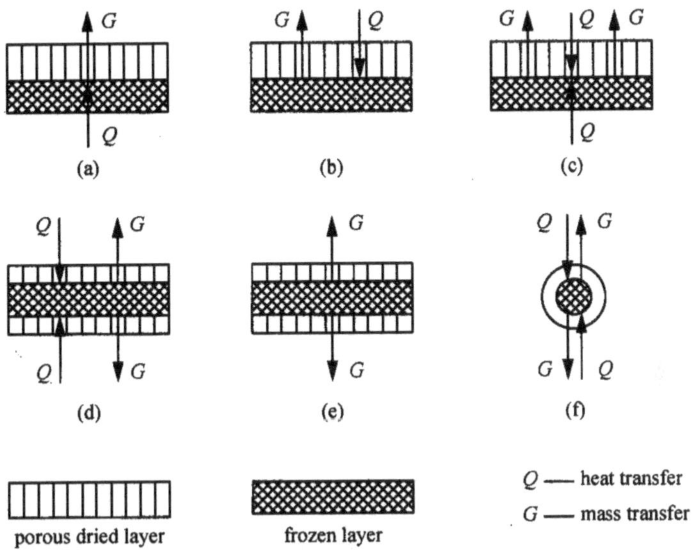

Fig.2.15 The common modes of heat transfer and mass transfer in the drying process.

(d) the heat rate Q is acquired from both the above porous dried layer and the below porous dried layer; the vapor rate G is also taken out from the two layers. There are two sublimation fronts in the mode. If the two sublimation fronts are moving at the same rate, the mode is symmetrical; otherwise, it is asymmetrical.

(e) the heat rate is acquired inside the material; the vapor sublimated rate G is taken out from the two dried layers. It is the case in microwave heating.

(f) it corresponds to the cases of sphere or columnar material.

2.3.3 Heat transfer

There are three modes in freeze drying: conduction, radiation and convection.

1. Heating by solid thermal conduction

The material is placed on a plate; and the plate is in direct thermal contact with the heating shelf, as the case in Fig. 2.15(a). The heat flow supplied by the heating shelf, passing through the plate and the frozen layer, arrives at the sublimation front. In this kind of freeze dryer, not only the high thermal conductivities of plate and shelf, but also the good contacts among the material, the plate and the shelf are required. Only in this way, the heat transfer can be enhanced, and the thermal uniform distribution inside the material can be achieved.

2. Heating by radiation and microwave

Thermal radiation is electromagnetic radiation emitted from the surface of an object which is due to the object's temperature. The radiation heat transfer is a form of thermal energy transition by electromagnetic wave, which may carry on in the vacuum. As the case in Fig. 2.15(d), the material held by a clamp is suspended inside the space between an upper shelf and a lower shelf. The heat flow supplied by both shelves is transmitted to the material by thermal radiation, and through the dried layers by thermal conduction, arrives the sublimation surface finally.

In this kind of freeze-drying equipment heating by radiation, good contact between the material and the shelf is not required. Therefore it is convenient for the realization of entire vehicle turnover through the drying chamber. This way of heating is widely used in large-scale freeze-drying equipment.

Microwave heating is accomplished by using microwave radiation to heat water and other polarized molecules within the material. Microwave heating is in essence the electromagnetic wave heating, except that the heating is directly produced within the material. No heat transfer from the exterior to the interior is needed, therefore the temperature at the center of the material is elevated quickly. Since microwave heating is hardly affected by the dried layer, the microwave heat flow can directly arrive at the sublimation front, thus the temperature difference between the outside surface of dried layer and the sublimation surface is very small, and it is possible to maintain the temperature of dried layer close to the highest temperature the material can bear.

3. Convective heating

Convective heat transfer is a mechanism of heat transfer arising from bulk motion of fluids. It plays a very small role in the drying step, for drying is carried out in vacuum condition. However a circulating pressure method is sometimes used to enhance the convective heat transfer to accelerate the drying. The method is to control the pressure of the drying chamber to undulate in certain scope thus to enhance the effects of heat transfer and mass transfer[47].

2.3.4 Mass transfer

During sublimation drying, there is massive vapor to be produced from the material. The vapor must be moved away promptly, otherwise the pressure in the drying chamber will be elevated, the sublimation surface temperature will increase, and finally the frozen layer will melt. Hence, a cold trap is set between the drying chamber and the vacuum pump. A cold trap is a device that condenses the water vapor into a liquid or solid (Another most common objective of cold trap is to prevent vapor from contaminating a vacuum pump). The water vapor escaped from the sublimation surface, passing through the dried layer, flows to the cold trap and forms frost on its surface. The driving force of the vapor flow is the vapor pressure difference between the sublimation surface and the cold trap. The main function of vacuum pump is to extract the non-condensable gas escaped from the material and the air leaked in from outside.

During the drying step, the temperature of the cold trap should be maintained at low enough temperatures in order to produce low enough vapor pressure. Regarding the freeze-drying of most materials, cold trap surface temperature within −40℃—−50℃ is applicable. But regarding certain materials with very low vitrification transition temperature T_g (or eutectic point temperature T_E), much lower cold trap surface temperature is required to guarantee the driving force of water vapor.

2.3.5 Moisture content distribution inside the material

Water content or moisture content is the quantity of water contained in a material, which is usually related to the mass of the material. There are two types of representations:

(1) the moisture content of the total material w: it refers to the water mass per unit mass of the material. It is also called the wet basis moisture content. The value of w expressed as $\frac{kg_{water}}{kg_{water}+kg_{dry}}$ can range from 0 (completely dry) to the value of the materials' porosity at saturation (<100%).

(2) the moisture content of the dried material m : it refers to the water mass per unit mass of the dry material. It is also called the dry basis moisture content. The value of m expressed as $\frac{kg_{water}}{kg_{dry}}$ may range from 0 to greater than 100%.

The two methods of reporting moisture content are related by the equations:

$$w = \frac{m}{m+1} \tag{2.27}$$

$$m = \frac{w}{1-w} \tag{2.28}$$

For example, a material has an initial moisture content of $w = 90\%$; its dry basis moisture content $m = 0.9/(1-0.9) = 9.0\ \frac{kg_{water}}{kg_{dry}}$. After the sublimation drying step, the wet basis moisture content w becomes 5%; its dry basis moisture content $m = 0.05/(1-0.05) = 0.526\ \frac{kg_{water}}{kg_{dry}}$.

In the ideal situation, the moisture content distribution within the material during the sublimation dry step is expressed as Fig. 2.16(a). There is an obvious interface between the frozen layer and the dried layer. The wet basis moisture content of the frozen layer is w_o ; and the wet basis moisture content of the dried layer is w_e. The moisture content suddenly drops across the interface. The ice sublimates into vapor on the interface, therefore the interface is called sublimation surface. As drying marches on, the sublimation surface moves gradually to the interior of the material (i.e. migrates to the left side in Fig.2.16).When the frozen layer thickness reaches zero, the sublimation surface vanishes completely, and the sublimation drying ends.

In real situation, the moisture content distribution within the material is like what is shown in Fig. 2.16(b). There is a certain moisture content gradient in the dried layer; moreover there is a transitional layer between the frozen layer and the dried layer, obvious moisture content gradient exits in the transitional layer. Although this layer does not contain ice crystal, the moisture content inside this layer is higher than the dry level remarkably.

Through a thorough research into the transitional layer, it is discovered that this layer is relatively thin, and the supposition of zero thickness does not cause an obvious error. Moreover, chemistry, physics and microscopic analyses have confirmed the point view: when the temperature of the frozen layer is lower than its vitrification transition temperature T_g, phenomena, such as material flowing, component redistribution as well as the structure destruction, will not appear in the transitional layer[55].

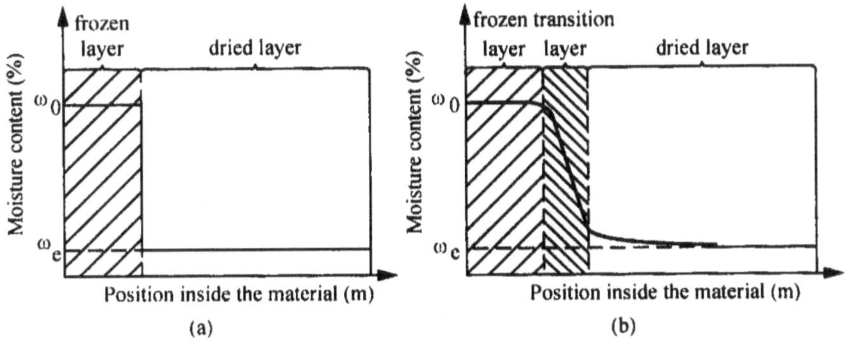

Fig.2.16 The moisture content distribution within the material during sublimation
(a) the ideal condition; (b) the actual condition

2.4 Desorption drying process

The desorption drying, also called secondary drying, is to heat the material under an elevated temperature higher than in the primary drying process, to break any physico-chemical interactions that have formed between the water molecules and the material, and to cause "the bound water" to become "free water", then evaporate to vapor. The heat needed in the desorption drying is the sum of the desorption heat and the evaporation heat, and "the desorption heat" is the shortened form of the sum.

From technological point of view, desorption drying is subsequent to the sublimation drying process; the frozen water has been removed completely in the sublimation drying. However, there may be some frozen water left in practice after sublimation drying, it will be melted, evaporated and carried out in the secondary drying process.

It is clear that the primary drying and the secondary drying have totally different mechanisms and operating parameters. Therefore, how to judge the time at which the sublimation dying ends and the desorption drying starts is very important. Too early or too late "switch" from the primary drying to the secondary drying would lead to poor quality of freeze-dried products, as well as waste of time and energy[48,49].

In the desorption process, it is not easy job to remove the moisture content absorbed within the dried layer. So it is necessary to elevate the heating temperature, reduce drying chamber pressure and decrease the cold trap temperature.

At the end of the secondary drying process, the moisture content within the material must meet certain requirement. Too high or too low residual moisture final (RMF) is disadvantageous. Higher water content will not do good to storage; lower water content may damage the activities of material. The final water content after desorption drying should be lower than 5% generally.

2.5 Storage of the freeze-drying products

The main purpose of freeze-drying is to strengthen the stability of products, i.e., to reduce the chemical and biological reactions to lengthen the shelf life of products from hours to days even years. However, even if the products are stable at the conclusion of freeze drying, they may also lose activeness in long-term storage. For example, the pancreatic proteinase having 100% activeness at the conclusion of freeze drying may lose 70% of its activeness if stored for two weeks at 40 ℃, relative humidity of 79%, and without protecting agent. The quality changes of products include protein condensation, deaminization, non-enzymatic browning (Maillard reaction), oxidation, hydrolysis and others. In order to guarantee the physical, chemical, microbiological and pharmacological natures of the products during the storing time, the products must go through a strict stability test procedure before they are pushed to the market.

2.5.1 Influence factors of product stability

1. Residual moisture content

The residual moisture content is one of the most important factors affecting stability of freeze-dried products. It is usually the controlling effect of protein stability. Water may participate directly in reactions as a reactant, or indirectly participate as a plasticizer or a response medium to affect the stability of freeze-dried products. Although the moisture content of the freeze dried product is very low, the enzymatic reaction can still occur if the enzyme activity has not been deactivated.

In addition, the water, as a plasticizer, may greatly reduce the vitrification transition temperature. From Gordon-Taylor equation, as the moisture content increases by 1%, the vitrification transition temperature T_g will drop 5—10K[16]. During storage, the freeze dried material is prone to absorb the moisture. If the vitrification temperature drops to below the storage temperature, stability of product will be weakened and may even result in collapse. In the storage period, the moisture content of the product may change because of many reasons, such as the leakage of packing material, the moisture content releasing from the rubber bottle stopper, the crystallization of amorphous ingredient and so on.

2. Storage temperature

Storage temperature is another most important factor affecting the stabilities of freeze-dried products. The temperature influence on solid preparation is complex. Generally speaking, the higher the temperature, the worse the stability is. Higher temperature will accelerate the protein condensation. When temperature is elevated, several phenomena may appear in the already freeze dried foods, pharmaceuticals and supplementary materials, such as melting, polymorphism transform and moisture content reduction[50].

Moreover, the high temperature can also speed up the chemical reaction and cause the degeneration of products. The freeze dried material has a loose structure, easy to rehydrate. The final resident moisture content of freeze dried material is often between 2%—5%; for some fruit juice food, it is lower than 2%.The corresponding vitrification transition temperature of these freeze dried material are usually above 0℃; and the collapse temperatures of most freeze dried materials are close or higher than the normal environment temperature.

When storage temperature is lower than the collapse temperature, the material loose structure can not be changed. However, if the material adsorbs the moisture slowly, its corresponding collapse temperature drops gradually. When collapse temperature is lower than storage temperature, several phenomena will appear, such as agglomeration, color deterioration and flavor change. The optimal storage temperature should be determined by the sugar ingredient and the resident moisture content of product. In general, lower temperature is always advantageous for long time storage.

3. Packing materials

Packing problem is often neglected by people. However if packing is not to be considered seriously, high quality end product cannot be obtained even if the product has the most stable prescription. The products stored in environment at room temperature will mainly be influenced by heat, light, water vapor and air (oxygen), the packing design is to insulate these influences. Simultaneously it must also consider the interaction between the packing material and the freeze dried product. The main packing materials widely used are glass, plastics, rubber and some metals. Some relative properties of them are listed in Table.2.5.

Table 2.5 Relative properties of some packing materials [50]

material	average density	Moisture permeance	gas permeance(O_2)	potential reactivity with product
Polyethylene (low density)	0.92	high	low	low
Polyethylene (high density)	0.96	low	low	low
Polypropylene	0.90	medium	low	low
Polyvinyl chloride (soft)	1.20	high	low	medium
Polyvinyl chloride (hard)	1.40	high	low	low
Polycarbonate	1.20	high	low	low
Polyamide (nylon)	1.10	high	low	high
Polystyrene	1.05	high	high	medium
PTFE	2.25	low	low	no
Sodium calcium glass	2.48	no	no	high
Borosilicate glass	2.23	no	no	low
Butyl rubber	1.30	low	medium	medium
Natural rubber	1.50	medium	medium	high
Chloroprene rubber	1.40	medium	medium	high
Polyisoprene rubber	1.30	medium	medium	medium
Silicone rubber	1.40	high	high	low

1) Glass containers

The physical and chemical properties of glass are stable. The glass is not permeable to water vapor and gases, and will not react with freeze dried products; therefore glass containers are the most widely used packing material at present. However glass has two shortcomings, namely, it is possible that alkaline matter is released and the glass fragment may fall off. Those are especially important to materials like injections. Brown glass can prevent penetration of light which wave length smaller than 470nm; therefore light sensitive pharmaceuticals are stored in the brown glass containers.

2) plastics

Plastics is a kind of high molecular polymer, including polyvinyl-chloride, polystyrene, polyethylene, polypropylene, polyester and gathers carbonic ether and others. For ease of forming or to prevent aging, plastics is frequently joined by some additives, plasticizer, and antioxidant. However some additives have toxicity, the plastic packing for medicinal purposes must use non-toxic plastic packing.

The plastic container has still some problems:

(1) The plastic container is permeable to gas. The product may exchange gas with the atmospheric gas, which may cause oxidation and volatility of the freeze dried products.

(2) The plastic container has some moisture permeability. For example, with temperature at 40℃ and relative humidity 90%, the polyvinyl-chloride film with 0.03mm thickness has moisture permeace factor of 100 $g/(m^2 \cdot d)$.

(3) The plastic container has also some adsorbability. Some ingredients in the product material (for instance, antiseptic) may be attracted by the plastics; simultaneously some ingredients of the plastic may also migrate to the product. For example, nylon can attract many kinds of bacteriostats. Choosing the right plastic material for packing is extremely important. High density polyethylene has high rigidity, skin hardness, and longitudinal strength. Its melting point and softening point are also high, and its permeablity to vapor and gas is very low. Therefore, high density polyethylene is widely used as packing material for freeze dried products.

3) rubber

Rubber is mainly used as the stopper or plug of the glass containers. It is widely applied at present. However the rubber stopper has also a series of problems which must be treated strictly.

In rubber forming process, some additives, such as vulcanizing agent, auxiliaries, antioxidant and others, have been added into the rubber. These chemical additives may pollute freeze dried products. Moreover, the rubber may also absorb the bacteriostat and some ingredients of the product. Specially, the adsorption of bacteriostat by rubber can cause reduction of bacteriostasis potency.

There is also some moisture in the rubber plug. The moisture will be released during storage process, which has certain influence on the quality of the products. The quantity of residual moisture inside the plug is mainly decided by the plug material and the treatment method. Grazio and the Flynn discovered that, the moisture content of a 215mg of freeze dried sucrose stored in a rubber plugged bottle for 3 months under room temperature increased from 1.95% to 2.65%[51].

Under an equilibrium state, the processes of moisture released from the rubber plug and captured by the product are relative to the mass and the moisture content of the plug, and the adsorption ability of the product.

Moreover the polymer, the chemical additive inside the plug can also affect the processes of absorption and desorption.

In order to stably preserve the quality of freeze dried products stored in bottle containers, the rubber plug must be as small as possible, and the plug material must also be optimized. The moisture content inside a rubber plug is relative to a great extent to the history which the plug has undergone. In general, the rubber plug having experienced steam sterilizing will capture 1.1% moisture content of its weight. The moisture content can fall to 0.1% after a vacuum drying for 8 hour, or by a hot air drying (110℃) for 8 hours.

The Fig.2.17 shows that the plug (curve 3), which undergoes steam sterilizing and 8 hours' vacuum drying, releases much less moisture to the lactose during storage than the plug (the curve 1) which undergoes a steam sterilizing without subsequent vacuum drying. When a plug undergoes a steam sterilizing but only 1 hour vacuum drying followed; the moisture content of the product, which is stored in a bottle with the said plug, may increase 2.4 times when it is stored at 25℃ for 6 months. It may also be observed from Fig.2.17 that the moisture content no longer changes after reaching the maximum value corresponding to the equilibrium state. The time required for reaching the equilibrium relies strongly on the storage temperature. There was a report for a given freeze dried product, the time for the moisture content reaching 1/2 of the maximum increased from 4 months to 10months when the storage temperature decreased from 40℃ to 5℃[3].

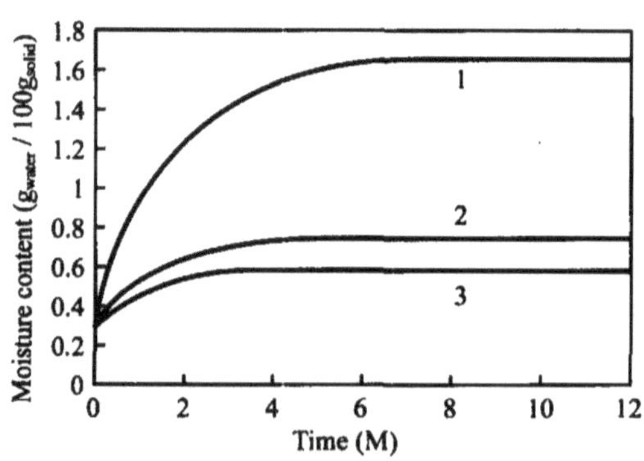

Fig.2.17 The relation between the moisture content of the freeze dried lactose and the storage time and the plug processing mode [3]

The rubber plugs with thickness of 13mm have undergone different processing modes:
1. steam sterilizing, without vacuum drying followed; 2. without either steam sterilizing or vacuum drying;
3. steam sterilizing, with 8 hours vacuum drying followed

Fig.2.18 gives the equilibrium (or the maximum) moisture content of lactose and vancomycin with different product mass and different treatment mode. Obviously, vancomycin captures moisture easier than lactose.

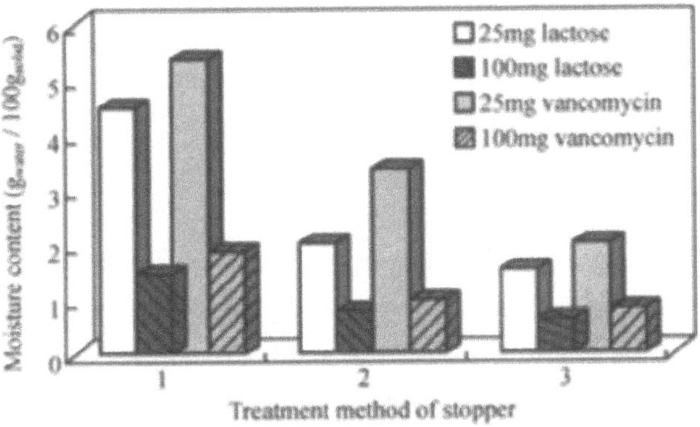

Fig.2.18 The relation between the maximum moisture content of the products and the product mass and the plug processing mode [3]

The rubble plugs with thickness of 13 mm have undergone different processing modes:
1. steam sterilized, without vacuum drying followed; 2. without either steam sterilizing or vacuum drying; 3. steam sterilized, with 8 hours vacuum drying followed

In storage, some unfavorable ingredients filled inside the bottle plug will diffuse to the products. Ordinary protection coating cannot prevent the penetration of these ingredients. However, Teflon may be used as a coating which has certain protective function.

In view of the fact that the packing material has strong influence on pharmaceutical stability, the packing testing of freeze dried product must be conducted carefully to choose the right packing material.

4. Glass transition temperature T_g

The glass transition temperature T_g may be used to judge the stability of freeze dried products. Generally speaking, the higher the value of T_g, the more stable the product is. The value of T_g is effected by the moisture content and the thermal history which the product has undergone. The valve of T_g can be measured by DSC (Differential Scanning Calorimetry)[4]. About how to choose the appropriate temperature for long-term storage, Frank believed the storage temperature should be at least 20K lower than its glass transition temperature T_g [52]. For example, if a freeze dried product is expected to be stored at 20℃ for long time, the glass transition temperature of the product should be higher than 45℃.

5. Others

1) Polymorphism in pharmaceutical

Polymorphism in materials science refers to the ability of a solid material to exist in more than one form of crystal' structure. Polymorphism describes different crystal packing arrangements of the same molecular species in its solid state, while the pseudo-polymorphs are the solvate and amorphous solid-state forms. Polymorphism investigations are particularly important in active drug and medical product development in the pharmaceutical industry, since certain properties of a formulated product are often directly related to the existing polymorphs in the formulation.

It has been found in the pharmaceutical production that the stabilities of a number of products are strongly affected by their polymorph forms. Many pharmaceutical products have several forms of polymorphs. For example, the rifampin (or rifampicin) has three forms of polymorph: amorphous, the form A and the form B. After an accelerated testing at 70℃for 15 days, more than 10% of the amorphous rifampin content inside the product is lost; the content is significantly decreased when stored at room temperature for 6 months. However only 1.5%~ 4% of the rifampin contents of the form I and the form II is lost after the said accelerated testing, and the contents inside the product are still higher than 90% after stored at room temperature for 6 months. Ampicillin sodium has also 3 forms of polymorphs: form A, form B and form C. The stability of form C is higher than form A and form B. Vitamin B1 also has several forms of polymorphs, mainly form A and form B. Form A is a water thiamine hydrochloride; and form B is a semi-crystalline water thiamine hydrochloride. The polymorph form A is very unstable in humid environment and prone to transit to form B, and form cakes and affect the quality of products.

2) Effect of the excipient

The medicinal products mainly consist of the active drug and the excipient. In freeze dried products, the excipient may have influence on product quality mainly because the moisture absorption of the excipient increases the moisture content of the product. For example, the reason why the poly three magnesium silicate is able to accelerate the aspirin degeneration is that it has very high moisture content. Certainly, the impact of the moisture on the quality of the medicinal product is relative to contact condition between the moisture and the active drug. Moreover, the interaction between the excipient and the active drug may also accelerate the degeneration of medicine.

3) The solid state of freeze-dried products

The active ingredients in freeze-dried products may exit both in crystalline state and in amorphous state. In general, the ingredients in crystalline state are more stable. For example, β-lactam antibiotics in crystalline state is more stable than in amorphous state.

4) The recrystalization of amorphous protective agents

Protective agents in freeze-dried products exit generally in amorphous state. The amorphous state is prone to change to crystal state since crystal state is more stable in thermodynamics[53]. Many sugar and polyol protective agents tend to recrystallize during storage. The rate of recrystallization increases with the temperature and the relative humidity[54]. The recrystallization would weaken the interaction between the protective agent and the protein, and reduce the glass transition temperature T_g , and make the product unstable.

In addition, there are other effects that have certain influence on the stability of freeze dried products, such as the concentration of solution, pH value, the illumination condition as well as the gas filled inside vessel.

2.5.2 Residual moisture content

1. The importance of controlling the moisture content

The purpose of freeze-drying is to maintain long-term biological activity of drug. The residual moisture content of the products is critical to stability of the active drug. The final requested residual moisture (FRM or RMF) of freeze-dried products is usually at 1.0%—3.0% or 5.0% (wt/wt). Too much or too little water content will affect the quality of freeze-dried products. The over drying will result in destruction of activity; and too much moisture will change the protein structure, reduce the drug activity and shorten the expiration date.

At the end of freeze drying, the final requested residual moisture is relative to the temperature, the chamber pressure and the time duration of drying. Since different excipient and lyoprotecant are added into the formula, the remaining moisture content inside the freeze dried products may also exist in different forms. Regarding moisture content, exerting control is of extreme importance. American Food and Drug Administration (FDA) stipulated that, the remaining moisture content of each batch of freeze dried product should be measured and conforms to the stipulative standard. Remaining moisture content measurements may have quite different results if different measurement techniques are used, even to the same product. Therefore, in order to compare the moisture content of products, same measurement technique must be used, and the data should also be obtained under same experimental condition.

2. Prevent re-absorption of moisture in the seal process

The balance moisture content of a freeze dried product is relative to the vapor pressure and the temperature in which the product locates. Their relationship may be indicated by a desorption isothermal as in Fig. 2.19. In the later period of drying process, excessively lengthened drying time is not necessary for it will not cause obvious decrease of the residual moisture content. Only at a higher temperature or a lower pressure, the residual moisture content can further be reduced.

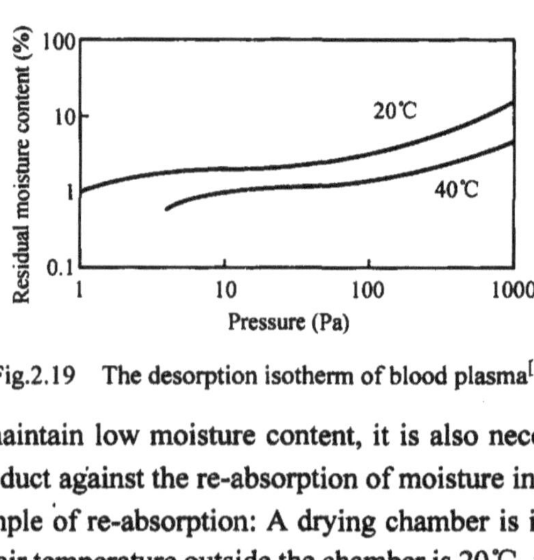

Fig.2.19 The desorption isotherm of blood plasma[3]

In order to maintain low moisture content, it is also necessary to prevent freeze-dried product against the re-absorption of moisture in the seal process. Here is an example of re-absorption: A drying chamber is in volume of 200 liters; the room air temperature outside the chamber is 20℃, relative humidity is 70% ; and the water content of unit air is approximately 1.15×10^{-2} g_{water} / L_{air}. If the room air leaks into the drying chamber for just a very short period of time, 2.3g water vapor is introduced into the drying chamber and absorbed by the product. By this way of re-absorption, the moisture of the product will increase obviously. If there are 300 vials in the chamber and each vial contains a product of 1 g, the balance moisture of the product will increase about 0.77% from the re-absorption.

For bulk drug material or food, the drying chamber has to be vented with dry air or inert gas. Water content of dry air should be very low. After filling products in vials or small bottles, the vials and bottles can be sealed in the drying chamber by a manipulators or stoppering systems.

3. Measurement methods of moisture content of freeze dried products

The residual moisture content of freeze-dried product should be tested. For the testing, the product must be handled to eliminate moisture absorption from the surrounding. For example, in order to prevent the moisture absorption, shifting product from one vessel to the other or weighing the product should be accomplished in a shielded box filled with inert gas or dry gas. The phosphorus pentoxide P_2O_5 is laid aside the product in the shielded box. Box should be handled with rubber gloves on. The balance used in the dry box has to be readjusted to avoid electrostatic charge. The test methods that have been used and approved are the following:

1) Gravimetric Method (Loss on Drying)

The principle of this method is using phosphorus pentoxide P_2O_5 as the desiccant to absorb water vapor from the product, to weight the product mass loss, then to calculate the moisture content of products. The measurement is done in a vacuum box, and the method is mainly used to measure the surface water and weakly bound water of the products.

The room air around the measurement device is kept at temperature of 20—30℃ and relative humidity of 20 to 25%. The mass of an empty bottle (or vial) m_0 is weighed; a freeze dried product is put in the bottle, the mass of the bottle with product m_1 is weighed. After the bottle has been dried inside a dryer for a few days (e.g. 3days), the mass of the bottle will be weighted again and again till the weight becomes constant, m_2. The moisture content of the tested product can be calculated by the following formula.

$$\text{Moisture content of the product} = \frac{m_1 - m_2}{m_1 - m_0} \times 100\% \qquad (2.29)$$

There is enough anhydrous phosphorus pentoxide P_2O_5 inside the dryer and the dryer box is vacuumed. The optimum sample size is about 200 milligrams. The temperature inside the dryer is 20℃ to 30℃. Higher temperature may accelerate the process but it may also cause the loss of volatile substances and damage of freeze dried products.

2) Karl Fischer Titration

The Karl Fischer titration was invented by German chemist Karl Fischer in 1935. The method is to measure the moisture content of gas, liquid and solid materials. A photo of Karl Fischer titrator is shown in Fig.2.20.

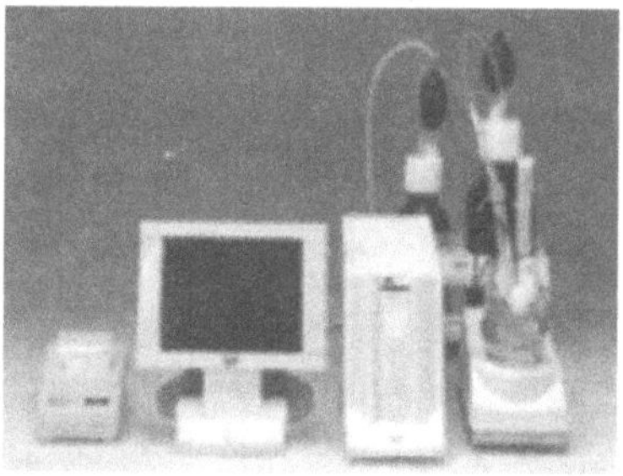

Fig.2.20 A Karl Fischer titrator (KEM Comp.)

The main compartment of the titration cell contains anode solution plus analyte. The anode solution consists of an alcohol (ROH), a base (*B*), SO_2 and I_2. A typical alcohol that may be used is methanol or diethylene glycol monomethyl ether with imidazole as a common base.

Titration cell also consists of a smaller compartment with an anode immersed in the anode solution of the main compartment. The two compartments are separated by an ion-permeable membrane. The Pt anode generates I_2 when current is provided through the electric circuit. The net reaction is oxidation of SO_2 by I_2. One mole of I_2 is consumed for each mole of H_2O. In other words, 2 moles of electrons are consumed per mole of water. Usually the end point is detected by bipotentiometric method. The amount of current needed to generate I_2 and to reach the end point can then be used to calculate the moisture content in the sample.

The volumetric titration is based on the same principles as the coulometric titration except that the anode solution above now is used as the titrant solution. The titrant consists of an alcohol (ROH), base (B), SO_2 and a known concentration of I_2. One mole of I_2 is consumed for each mole of H_2O.

The advantages of Karl Fischer titration method over conventional loss on drying (LOD) thermal methods of moisture determination are its high accuracy, small sample quantities requirement, and short duration for analysis. Most importantly, Karl Fischer titration method has the specificity for water, but the gravimetric method will detect the loss of any volatile substance (including water and others)[147].

However the Karl Fischer methodology cannot be used in the following cases: The product material interferes with Karl Fischer reagents; the sample does not adequately dissolve in the Karl Fischer reagent (e.g. methanol) ; the sample moisture does not adequately extracted into the solvent. Other methods should be used to determine residual moisture for these cases.

3) Thermogravimetry

Thermogravimetry (TG) is based on the continuous recording of mass changes of a sample, as a function of temperature and time. A sample is placed on an arm of a recording microbalance, where that arm and the sample are placed in a furnace. The furnace temperature is controlled in a pre-programmed temperature/time profile. The photo of a thermogravimeter is shown in Fig.2.21.

Fig.2.21 A thermogravimeter (Pyris 1, Perkin Elmer Comp.)

Other alternate methodologies to measure the residual moisture content include gas chromatography, moisture content analyzer, near infrared light spectral method, the tritium isotope method etc. Those methods are used less frequently than the previous three methods.

Water in freeze dried product can be bound in several different forms: as surface water, as water bound more or less to the dried substance, or as water of crystallization or vitrification state. Therefore each method of measurement can lead to different results for different substances. Moisture contents of several substances measured by different methods are shown in Tab.2.6. The residual moisture content (RM) measured by gravimetry are about 0.3%—0.6% of RM lower than that by the Karl Fischer titration, while the values of RM by thermogravimetry are close to that by Karl Fischer titration[3].

Table 2.6 Comparison of the residual moisture content (RM) of vaccines measured by different methods

testing method	residual moisture content rm (%±standard deviation) in the vaccine			
	rubella virus	mumps virus	rubella and mumps virus	measles, mumps and rubella virus
gravimetry	0.42±0.18	1.10±0.40	0.41±0.26	0.18±0.14
karl fischer titration	1.03±0.14	1.54±0.20	0.72±0.16	0.80±0.14
thermogravimetry (hold at 60℃)	1.17±0.20	1.53±0.17	0.74±0.13	0.70±0.08

2.5.3 The testing methods of lyophilized product stability

The purpose of stability testing is to provide evidence on how the quality of lyophilized product varies with time under the influence of a variety of environmental factors, such as temperature, humidity, and light, and to establish a shelf life for the product and recommended storage conditions.

1) Shelf life (also referred to as expiration dating period)

Lyophilized product is considered stable as long as its activity or its potency remains within a given range of values. For example, the potency of an active constituent may range from 110% to 90%. The shelf life is the length of time that all of the lyophilized formulation under the condition as shown in the label remains within the specified potency limits. To determinate the expiration date, long-term stability testing is necessary.

2) Long term testing (real-time testing)

Long term testing is a stability study under recommended storage condition for the re-test period or shelf life proposed (or approved) for labeling. The biological activity is generally reported at discrete time intervals.

Among testing methods of predicting product stability, the long term testing at room temperature is the best one that accurately reflects the actual situation, and is the best basis to predict the expiration date. However, the drawbacks of long-term stability testing are : very long time is needed (from months to years, for example four years) it is not easy to detect and correct problems in time.

Long-term testing is carried out in the actual storage conditions. For example, the test products can be divided into three batches, packaged as on the market, placed for 12 months. The temperature is (25±2)℃, the relative humidity is (60±5)%.

For long-term studies, frequency of testing should be sufficient to establish the stability profile of the product. For drug substances with a proposed retest period of at least 12 months, the frequency of testing at the long-term storage condition should normally be every 3 months over the first year, every 6months over the second year, and annually thereafter through the proposed retest period. If the test time is not long enough (for example, only 18 months) to determinate the expiration date, statistical analysis may be used to determine the validity of the product. If statistical data of those three batches are quite close, the average of them can be chosen for determining the expiration date; if there is large difference among the three batches of data, the shortest data will be chosen to determine the expiration date. More about the statistical analysis for determining the expiration date may be found in the pharmacopoeia such as the Chinese Pharmacopoeia 2000 version.

Lyophilized products which are particularly sensitive to the temperature must be stored in a refrigerator. Long-term testing of such products is carried out at the temperature of (5±3)℃ for 12 months. Samples are taken for stability testing every 3 months over the first year. If the testing is to be taken after 12 months, the continual inspection is still needed.

3) Accelerated testing

Accelerated testing is carried out in extraordinary conditions. The aim is to increase the rate of chemical degradation or physical change of lyophilized products by using exaggerated storage conditions (e.g. elevated temperature or strong illumination) as part of the formal stability studies, in order to provide necessary information to products packaging, transportation and storage.

In accelerated testing, temperature is set at about (40±2)℃, the relative humidity RH is (75±5)%, and the minimum time period covered by data at submission is 6 months. Alternative storage conditions can be used if justified.

Intermediate testing is designed to moderately increase the rate of chemical degradation or physical changes for a drug substance or drug product intended to be stored long-term at 25°C. The intermediate testing conditions are: (30±2)℃; RH: (65±5)% and the minimum time period covered by data at submission is 6 months.

For the drug substances intended for storage in a refrigerator, the accelerated testing conditions are: (25±2)℃; the relative humidity RH (60±5)%, and the minimum time period covered by data at submission is 6 months.

The shelf life of lyophilized product can be inferred from the data of accelerated testing, the method of which can be found in reference [2] or [4].

3

Heat-Mass Transfer Analyses and Modeling of the Drying Process

3.1 Restriction of heat-mass transfer and its relative properties

In order to keep sublimation drying going smoothly, two basic conditions must be satisfied: vapor produced from sublimation must be moved away unceasingly from the sublimation surface and heat required by sublimation must be supplied unceasingly by a heater. As a matter of fact, sublimation drying is a process in which heat transfer and mass transfer carry on simultaneously, whereas the heat transferred to the sublimation surface must equal the heat with which the vapor escaped from the sublimation surface.

3.1.1 The restriction of mass transfer

The mass transfer process inside the material during sublimation is to be discussed firstly. Let's denote the temperature of cold trap as T_a, its corresponding saturation vapor pressure as P_a; the sublimation interface temperature as T_i, its corresponding saturation vapor pressure as P_{iw}.

The driving force of mass transfer process is $(P_{iw} - P_a)$; and the resistance of mass transfer is composed of three parts: the resistance inside the dried layer R_d, the resistance from the surface of dried layer to the surface of cold trap R_s, and the sublimation resistance of solid ice R_I.

Thus, the sublimation rate of ice G can be expressed as

$$G = \frac{p_{iw} - p_a}{R_d + R_s + R_I} \tag{3.1}$$

where G ——the sublimation rate of ice, $\mathrm{kg/(m^2 \cdot s)}$;

R_d ——the resistance inside the dried layer, $\mathrm{m^2 \cdot Pa \cdot s/kg}$;

R_s ——the resistance from the surface of dried layer to the surface of cold trap, $\mathrm{m^2 \cdot Pa \cdot s/kg}$;

R_1 ——the sublimation resistance of solid ice, $R_1 = \frac{\sqrt{T_i}}{K_1}$, K_1 is a constant relative to the molecular weight of the sublimation material.

For pure ice,

$$K_1 = 0.018 \cdot \text{kg} \cdot \text{K}^{1/2} / (\text{m}^2 \cdot \text{Pa} \cdot \text{s})$$

Now discuss the mass transfer process of pure ice sublimation as an extreme situation. The surface temperature of the pure ice is indicated as T_i, its corresponding saturated vapor pressure as p_{iw}. The cold trap temperature T_a is very low, its corresponding saturated vapor pressure P_a is much lower than p_{iw}, that is $P_a << P_{iw}$. The sublimation surface is the ice surface itself, the resistance $R_d = 0$. Further assume that the flow resistance from the surface of pure ice to the surface of cold trap R_s can be neglected; at this point, the sublimation rate reaches the maximum G_{max} [55].

$$G_{max} = \frac{p_{iw}}{R_1} = \frac{K_1 p_{iw}}{\sqrt{T_i}} \tag{3.2}$$

where G_{max} ——the maximum sublimation rate of pure ice, $\text{kg}/(\text{m}^2\text{s})$;

T_i——The temperature of pure ice, K;

p_{iw} ——the saturated vapor pressure of pure ice, Pa;

$$K_1 = 0.018 \cdot \text{kg} \cdot \text{K}^{1/2} / (\text{m}^2 \cdot \text{Pa} \cdot \text{s}).$$

For pure ice at -10 ℃, its corresponding saturated vapor pressure is 260 Pa

$$G_{max} = \frac{0.018 \times 260}{\sqrt{263}} = 0.29 \ \text{kg}/(\text{m}^2 \cdot \text{s}) = 1044 \ \text{kg}/(\text{m}^2 \cdot \text{h})$$

In order to fulfill the maximum sublimation rate G_{max}, the heat flux required is

$$Q_{max} = 0.29 \times 2836 = 822.44 \ \text{kW}/\text{m}^2$$

In fact the actual sublimation drying rates are only within the range of 1—10 $\text{kg}/(\text{m}^2 \cdot \text{h})$, much lower than the maximum rate of sublimation. On one hand, flow resistance inside the dried layer R_d and that from the dried layer surface to the cold trap R_s coexist. On the other hand the actual heat transfer cannot supply heat flux high enough.

3.1.2 Restrictions of heat transfer

1. Temperature limit

Sublimation requires heat supply to the material. No matter what the heating methods is, the materials will be subject to the following types of temperature restrictions. Firstly, if a eutectic solid is formed during freezing, the heating temperature during sublimation must be lower than its eutectic temperature of the frozen layer to prevent it from melting; if a glassy solid is formed during cooling, the heating temperature must be lower than its glass transition temperature to avoid devitrification. Secondly, temperature of the dried layer has to be lower than its collapse temperature to avoid its collapse. Thirdly, surface temperature of the dried layer should be lower than its permissible temperature to avoid scorch or losing activeness.

Table 3.1 and Table 3.2 list the maximum surface temperatures, the collapse temperature, the glass transition temperature and other characteristic temperatures of some materials.

Table 3.1 Maximum temperatures of dried layer and frozen layer of some materials in radiation heating cases [56]

material	drying chamber pressure (Pa)	the maximum surface temperature of dried layer (℃)	temperature of frozen layer (℃)
chicken pieces		60	-20
strawberry chip	60.0	70	-15
orange juice	6.7—13.3	49	-43
Guava juice	6.7—13.3	43	-37
shrimp	13.3	52	-29
salmon fillets	13.3	79	-29
beef (rapid frozen)	66.7	60	-14
beef (slow freezing)	66.7	60	-17

Table 3.2 Melting temperatures, collapse temperature and glass transition temperature of some carbohydrates [16]

carbohydrate	characteristics of dried material				characteristics of aqueous solution	
	molecular weight	T_m (℃)	T_g (℃)	T_m/T_g	$T_g' \approx T_c \approx T_r$ (℃)	W_g' (wt%)
glycerol	92.1	18	-93	1.62	-65	46
xylose	150.1	153	9~14	1.49	-48	31
ribose	150.1	87	-10~-13	1.37	-47	33
glucose	180.2	158	31~39	1.39	-43	29
fructose	180.2	124	7~17	1.39	-42	49
galactose	180.2	170	30~32	1.45	-41~-42	29~45
sorbitol	182.2	111	-2~-4	1.45	-43~-44	19
sucrose	342.3	192	52~70	1.40	-32~-46	20~36
maltose	342.3	129	43~95	1.19	-30~-41	20
trehalose	342.3	203	77~79	1.35	-27~-30	17

continued

carbohydrate	characteristics of dried material				characteristics of aqueous solution	
	molecular weight	T_m (℃)	T_g (℃)	T_m/T_g	$T_g' \approx T_c \approx T_r$ (℃)	W_g' (wt%)
lactose	342.3	214	101	1.37	-28	41
maltotriose	504.5	134	76	1.17	-23~-24	31
maltopentose	828.9		125~165		-15~-18	24~32
maltohexose	990.9		134~175		-14~-15	24~33
maltoheptose	1153.0		139		-13~-18	21~33

T_m — the melting temperature;

T_g — the glass transition temperature;

T_g' —the glass transition temperature at maximally frozen- concentrated state;

T_c — the collapse temperature;

T_r — the recrystalization temperature;

W_g' — the water content of solution at maximally frozen- concentrated state.

2. Resriction of thermal conduction

Due to the limit on temperature, the temperatue difference and the thermal conduction are both very small during sublimation drying. As an example, a frozen material is heated by thermal conduction, as in Fig.2.15(a). Suppose the frozen layer has a thickness of 1cm; and the temperature difference across the layer is 10K. Although ice has high thermal conductivity, K_f =2.24 W/(m·K), the heat transfer flux through the frozen layer is still very low. That is,

$$Q = \frac{K_f \cdot (T_{lw} - T_i)}{x_f} = 2.24\text{kW/m}^2 \tag{3.3}$$

where Q—— the heat flux through the frozen layer, W/m^2;

K_f —— the thermal conductivity of ice, W/(m·K);

T_{lw} —— the bottom surface temperature of the frozen layer, K;

T_i —— the sublimation surface temperature, K;

x_f —— the thickness of frozen layer, m.

The heat flux (2.24kW/m^2) is only 0.27 % of the heat flux 822.44 kW/m^2 for the maximum sublimation rate as shown in Equation (3-3); and it can only supply a sublimation rate of 2.24 /2836 = 0.00079 $\text{kg/(m}^2\cdot\text{s)}$ = 2.75 $\text{kg/(m}^2\cdot\text{h)}$.

3. Resriction of thermal radiation

The heat flux Q is input by thermal radiation through the top of porous dried layer, as shown in Fig.2.15(b).

$$Q=\sigma\cdot F_{1-2}\cdot\left(T_{up}^{4}-T_{s}^{4}\right) \tag{3.4}$$

where Q——the heat flux by thermal radiation, W/m^2;

σ——Stefan-Boltzmann constant, namely black body radiation constant, $5.669\times10^{-8}\,W/(m^2\cdot K^4)$;

F_{1-2}——the shape factor;

T_{up}——the temperature of radiation heating plate, K;

T_s——the surface temperature of dried layer, K.

As a result of the material temperature limitation, the temperature of radiation heater cannot be too high. Support $T_{up}=313K$, T_s=263K and F_{1-2} = 1.0, the heat flux of thermal radiation is

$$Q=5.67\times10^{8}(T_{up}^{4}-T_{s}^{4})=0.27\ (kW/m^2)$$

In this case, the heat flux from thermal radiation $0.27\cdot kW/m^2$ is even smaller than that by thermal conduction; and the heat flux can only supply the sublimation rate of $0.34\,kg/(m^2\cdot h)$.

4. microwave heating and its restriction

Microwave heating is also electromagnetic heating. The frequency of microwave radiation is between the common radio and the infrared frequencies. The household microwave oven has frequency of 2450 MHz; the common large industrial microwave oven has frequency of 915 MHz . The microwave heating works by passing non-ionizing microwave radiation through the materials. Water, fat, and other substances in the foods and pharmaceuticals can absorb energy from the microwaves in a process called dielectric heating. Many molecules (such as water molecules) are electric dipoles, meaning that they have a positive charge at one end and a negative charge at the other, and therefore rotate as they try to align themselves with the alternating electric field of the microwaves. This molecular movement creates heat as the rotating molecules hit other molecules and put them into motion.

The microwaves can penetrate the material more deeply than with other methods; the depth of initial heat deposition may be several centimetres or more, depending on the water content. The microwave can penetrate the dried layer and reach the sublimation surface. The temperature diffrence between the surface of dried layer and the sublimation surface is very small. Therefore the sublimation temperature can be elevated to close to the material permission maximum temperature.

The heating rate of microwave per volume can be calculated by

$$P = 55\times10^{-14} E^2 \cdot \nu \cdot \varepsilon \tag{3.5}$$

where P—the heating rate of microwave per volume, $\mathrm{W/cm^3}$;

E— the electric field strength, $\mathrm{V/cm}$;

ν— the frequency of microwave, Hz;

ε — the microwave absorption coefficient of material.

The microwave absorption coefficient of material ε represents a property of material. It is closely related with the components, temperature and state of the material. Table 3.3 lists the microwave absorption coefficient ε of beef at different temperatures and states(2450 MHz). Microwave heating is more efficient on liquid water than on fats, sugars (which have less molecular dipole moment) and frozen water (where the molecules are not free to rotate).

Table 3.3 The microwave absorption coefficient ε of beef at different temperatures and states(2450 MHz)

temperature	beef	freeze dried beef
-40℃	0.083(frozen)	0.0058
-18℃	0.293(frozen)	0.0079
4℃	10.56(unfrozen)	0.0122

For microwave of 2450 MHz, the microwave absorption coefficient ε of materials containing frozen water is much smaller than that containing unfrozen water in 1-2 orders of magnitude and the values of ε of freeze dried materials are lower than the frozen materials in another order of magnitude.

Theoretically, microwave heating does not have the above restrictions of thermal conduction and thermal radiation, and may arrive at very high heating rate and sublimation rate, however microwave heating has only obtained little applications up to now. The main reason is that the components and the water contents are not uniform inside the materials and the microwave heating is extremely difficult to control. The microwave absorption coefficients may vary in several orders of magnitude at different parts inside the material. Thus, the sublimation process may be destroyed and lead to undesirable phenomena such as scorch created in some spots of the material while the majority of the material is still frozen. In addition, under the effect of strong electric field in microwave, the gas in drying chamber may be ionized, then the structure and qualities of the material may also change.

3.2 Characteristics of drying process in different heating forms

3.2.1 Characteristics of drying process where heating is from radiation through dried layer(s)

1. Analysis modeling

Let's take a thin slice of material as an example, to analyze the radiation heating from both sides. This is a case of heat transfer inward from two-side surfaces and mass transfer outward from two-side surfaces, as shown in Fig. 2.15(d).

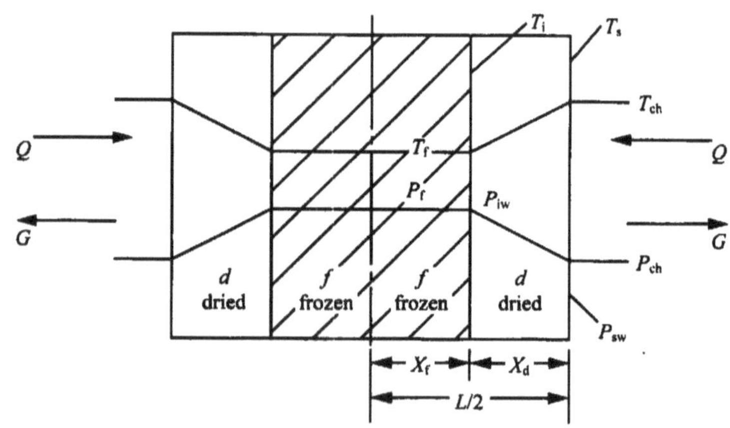

Fig.3.1 The analysis sketch for primary drying of plate material heated by radiation from two sides

Fig.3.1 is a schematic drawing to analyze the sublimation drying where material is heated by radiation from two sides. Let's denote the thickness of material as L, the thickness of dried layer as x_d, its temperature as T_d ; the thickness of frozen layer as x_f, its temperature as T_f; the temperature at sublimation interface as T_i; the temperature and the vapor pressure at outside surface of dried layer as T_s and P_{sw} respectively; the vapor pressure and the temperature inside drying chamber as P_{ch} and T_{ch}, respectively.

Assume that the vapor pressure at the interface P_{iw} is the saturation pressure corresponding to temperature T_i; the outside surface temperature of the dried layer T_s has reached the highest permissible temperature, and is maintained unchanged during the sublimation drying; the vapor pressure at the outside surface of the dried layer P_{sw} is lower than the saturation pressure $P_{sat}(T_s)$ of temperature T_s; the drying chamber pressure P_{ch} is lower than the saturation pressure of temperature T_{ch}; the vapor pressure P_{sw} at the outside surface of dried layer approximates the drying chamber pressure P_{ch}, which is maintained unchanged.

Radiation heat arrives firstly at the outside surface, then passes through the dried layer, and finally arrives at the sublimation interface, as shown in Fig.3.1.

Regarding such cases, heat flux Q and sublimation rate G are related to the temperature T_i and its corresponding vapor pressure P_{iw} at the sublimation interface. How to determine T_i or P_{iw} will be discussed below.

2. Determination of the sublimation temperature T_i and pressure P_{iw}

The heat transfer inside the dried layer is performed by thermal conduction only. The heat flux Q, $\mathrm{W/m^2}$, is

$$Q = \frac{K_d}{x_d}(T_s - T_i) \tag{3.6}$$

where x_d——the thickness of the dried layer, m;

K_d——the thermal conductivity of the dried layer, $\mathrm{W/(m \cdot K)}$.

The sublimation rate per unit area G, $\mathrm{kg_w/(h \cdot m^2)}$, depends on the mass transfer by diffusion. That is

$$G = -\frac{\mathrm{d}W}{\mathrm{d}t} = b(P_{iw} - P_{sw})/x_d \tag{3.7}$$

where W——the total moisture contain inside the material per unit area, $\mathrm{kg_w/m^2}$;

t——the drying time, h;

b——the water vapor permeability through dried layer, $\mathrm{kg_w/(m \cdot Pa \cdot h)}$.

In sublimation process, the sublimation surface is an interface between a dried porous layer and a frozen layer. The interface is moving from the outside surface into the center of the material. By the mass-conservation equation, there is

$$G = -\frac{dW}{dt} = \rho(m_f - m_d)\frac{dx_d}{dt} \tag{3.8}$$

where m_f ——moisture content per unit mass of dry material inside the frozen layer, kg_w / kg_d;

m_d ——moisture content per unit mass of dry material inside the dried layer, kg_w / kg_d;

ρ ——density of the dry material, kg_d / m^3.

By the energy conservation equation, there is

$$Q = G \cdot \Delta H_s \tag{3.9}$$

where ΔH_s ——the sublimation heat per unit ice, kJ / kg_w.

By introducing Eq. (3.6) and Eq.(3.7) into Eq. (3.9), we have

$$\left.\begin{aligned} \frac{b(P_{iw} - P_{sw})}{x_d}\Delta H_s &= K_d(T_s - T_i)/x_d \\ P_{iw} &= P_{sw} + \frac{K_d}{b\Delta H_s}(T_s - T_i) \end{aligned}\right\} \tag{3.10}$$

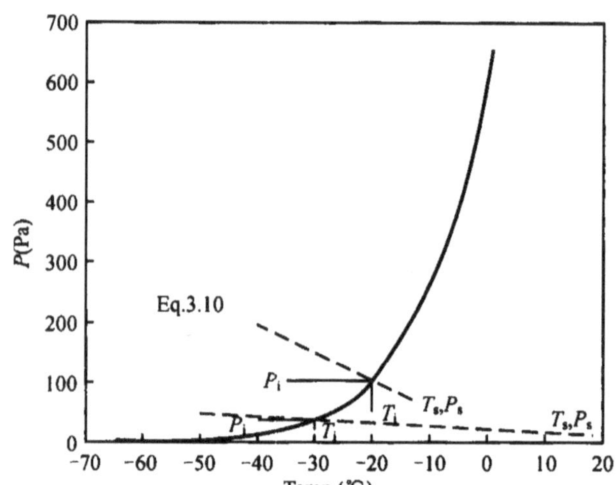

Fig.3.2 Determination of sublimation temperature T_i and pressure P_{iw} in drying process by radiation heating

The outer surface temperature T_s and pressure P_{sw} of the dried layers remian unchanged

Because pressure P_{iw} is the saturated pressure of temperature T_i, the relation of P_{iw} and T_i is definite, as shown in Table 2.3. Combining the said relation and Eq. (3.10), the values of sublimation interface temperature and pressure can be determined as shown in Fig.3.2.

In a sublimation drying stage, the thickness of dried layer x_d keeps increasing , and the thickness of frozen layer x_f keeps decreasing. However, as long as the outer surface temperature T_s remains unchanged, the sublimation interface temperature T_i and pressure P_{iw} will remain unchanged till the end of primary drying stage.

3. Determination of the time t_d for sublimation drying

Combining Eq.(3.7) and Eq.(3.8) gives

$$\rho(m_f - m_d)\frac{dx_d}{dt} = b(P_{iw} - P_{sw})/x_d \tag{3.11}$$

Integrate both sides of the equation

$$\int_0^{L/2} x_d dx_d = \frac{b(P_{iw} - P_{sw})}{\rho(m_f - m_d)} \int_0^{t_d} dtx$$

The result is

$$t_d = \frac{L^2 \rho(m_f - m_d)}{8b(P_{iw} - P_{sw})} \tag{3.12}$$

Seemingly, the solution of t_d has no relations with the thermal conductivity K_d and radiation heat transfer Q. However in fact, the saturated pressure P_{iw} at the sublimation intersurface in Eq.(3.12) is determined by Eq.(3.10), which has been taken into account the heat transfer .

3.2.2 Characteristics analyses of sublimation process where heat is from conduction through the frozen layer

1. Analysis modeling

Let's take a thin slice of material as an example to analyze the conductive heating from one side. This is a case of heat transfer inward from one side surface, and mass transfer outward from another side surface, as shown in Fig. 2.15(a).

Fig.3.3 is a schematic drawing to analyze sublimation drying where material is heated by thermal conduction from one side. Let's denote the temperature of lower heating plate as T_{lp}, the temperature of lower surface of material as T_{lw}. The control equations of heat and mass transfer are similar to that in the above paragraph. However, the characteristics of this case are: the heat conduction equation pertains to the frozen layer and the mass transfer equation pertains to the dried layer. Assume both the lower surface temperature T_{lw} and the outside surface vapor pressure P_{sw} are constant during the sublimation drying, the heating flux Q (W/m^2) can be expressed as

$$Q = \frac{K_f}{x_f}(T_{lw} - T_l) \tag{3.13}$$

where x_f —the thickness of the frozen layer, m;

K_f —the thermal conductivity of the frozen layer, $W/(m \cdot K)$.

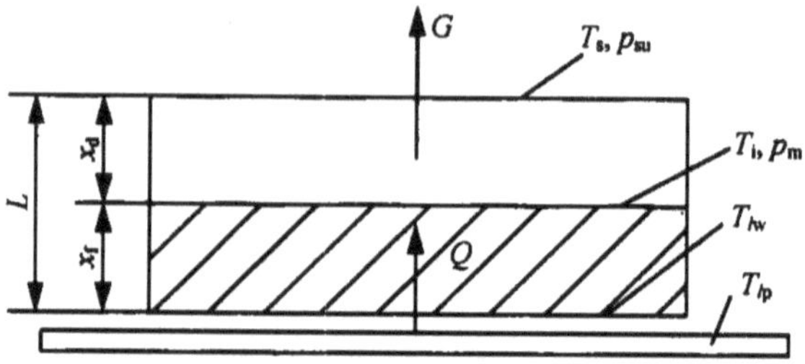

Fig.3.3 The analysis sketch for primary drying of plate material heated by thermal conduction from one side

The sublimation rate G, $kg_w/(h \cdot m^2)$, depends on the mass transfer by diffusion. That is

$$G = -\frac{dW}{dt} = b(P_{iw} - P_{sw})/x_d \tag{3.14}$$

where W —the total moisture contain inside the material per unit area, kg_w/m^2;

t —the drying time, h;

b —the water vapor permeanbility through the dried layer, $kg_w/(m \cdot Pa \cdot h)$.

2. Determination of the parameters of the sublimation surface (T_i or P_{iw})

Introducing Eq.(3.13) and Eq.(3.14) into Eq.(3.9) gives

$$P_{iw} = P_{sw} + (\frac{K_f}{b\Delta H_s})(\frac{x_d}{x_f})(T_{lw} - T_i) \tag{3.15}$$

where x_d and x_f are the thickness of dried layer and frozen layer, respectively.

Obviously,

$$x_f + x_d = L \tag{3.16}$$

Comparing Eq. (3.15) with Eq. (3.10), the right side of Eq. (3.15) contains K_{f}, but not K_{d}; moreover it has $\left(\frac{x_{\mathrm{d}}}{x_{\mathrm{f}}}\right)$. In the case of one-side conduction heating sublimation drying, the sublimation surface temperature T_{i} (as well as its corresponding vapor pressure P_{iw}) is no longer fixed; it is function of the dried layer thickness x_{d}. That is

$$T_{\mathrm{i}}(\text{or } P_{\mathrm{iw}}) = Fun \cdot (x_{\mathrm{d}}) \tag{3.17}$$

This function can be determined by Eq. (3.15) and the water vapor pressure-temperature relation (Table 2.3). Its operating procedure is as below: set a value of x_{d}, then draw a line according to Eq. (3.15); according to the method similar to that in Fig.3.2, determine the value of T_{i} (and P_{iw}); and then set up other values of x_{d}, determinate their corresponding T_{i} (and P_{iw}).

3. Determination of the time t_{d} for sublimation drying

In sublimation process, the dried layer thickness x_{d} is increasing. From the mass conservation equation, there is

$$G = -\frac{\mathrm{d}W}{\mathrm{d}t} = \rho(m_{\mathrm{f}} - m_{\mathrm{d}})\frac{\mathrm{d}x_{\mathrm{d}}}{\mathrm{d}t} \tag{3.18}$$

Combining Eq.(3.14) and Eq. (3.18) gives

$$\rho(m_{\mathrm{f}} - m_{\mathrm{d}})\frac{\mathrm{d}x_{\mathrm{d}}}{\mathrm{d}t} = b(P_{\mathrm{iw}} - P_{\mathrm{sw}})/x_{\mathrm{d}} \tag{3.19}$$

Eq. (3.19) appears to be the same as Eq. (3.11), in fact the vapor pressure P_{iw} of the sublimation intersurface in Eq. (3.19) is no longer constant; it is a function of x_{d}, that is $P_{\mathrm{iw}} = Fun(x_{\mathrm{d}})$. Therefore the integral becomes

$$\int_0^L \frac{x_{\mathrm{d}}}{P_{\mathrm{iw}} - P_{\mathrm{sw}}}\mathrm{d}x_{\mathrm{d}} = \frac{b}{\rho(m_{\mathrm{f}} - m_{\mathrm{d}})}\int_0^{t_{\mathrm{d}}} \mathrm{d}t \tag{3.20}$$

Regarding such sublimation drying case where material is heated by thermal conduction from one-side, the resistance of heat conduction is small since the frozen layer has very high thermal conductivity (higher than the dried layer approximately in 1—2 orders of magnitude). The resistance of mass transfer in dried layer is also small in the initial stage of sublimation, therefore the sublimation rate is also high. As the dried layer is thickening, the mass transfer resistance increases and the sublimation process becomes slower.

If the upper part of dried layer can be removed unceasingly, the resistance of mass transfer will be reduced and the sublimation process will speed up and less time is needed for sublimation.

3.3 Analyses of mass transfer control and heat transfer control in sublimation drying

Sublimation drying is a process with interface moving, in which heat transfer and mass transfer take place simultaneously. There are many factors affecting the drying process which make heat transfer and mass transfer more complicated. These factors include heat transfer mode, drying chamber pressure, process parameters and the materials parameters.

Several mathematical models have been proposed to describe sublimation drying process, the most widely used model is the uniformly retreating ice front model (referred to as URIF model) proposed by King[57].

The URIF model is established on the supposition that the sublimation surface temperature, the material surface temperature and the vapor pressure inside dried layer are all constant. Other assumptions of URIF model mainly include: ①there is an obvious ice front between the frozen layer and the dried layer, and the thickness of sublimation interface can be ignored; ② the ice front retreats uniformly; ③ the non-ice layer formed is an absolutely dry porous layer.

This model has been applied in calculation of the initial 60%—90% moisture content removal, the moisture content of which is free water. In other words, URIF model can only be used to explain the sublimation stage. Since this model is relatively simple and can be used to estimate the sublimation rate, many researchers use it to describe the sublimation process and to predict the drying time of some frozen drying materials or to establish the operating policy.

In sublimation drying, if the heat supplied to sublimation surface equals to the heat required by the sublimation, the temperature and pressure of sublimation surface become stable and the sublimation process carries on normally. However, if the heat supplied to the sublimation surface is insufficient, the sublimation speed will drop; if the vapor diffusion resistance in the dried layer is too large, the sublimation surface pressure and the temperature may rise, and may result in the melting of frozen material.

If the mass transfer conditions are excellent, while the heat transfer conditions are relatively poor, the heat transfer conditions become the principal drawback, this kind of sublimation process is called a heat transfer controlled process. Otherwise, if the heat transfer conditions are excellent while the mass transfer conditions are relatively poor, this kind of sublimation process is called a mass transfer controlled process.

Because of the diversity of materials and processing parameters, it is very difficult to distinguish between the heat transfer controlled process and the mass transfer controlled process. However, generally speaking, the heat transfer obstacle does not appear in cases where heating is conducted through the frozen layer and the mass transfer obstacle does not appear in the initial stage of sublimation drying. In many actual sublimation processes, and for most of the drying time, using mass transfer control model for analysis may be more reasonable.

3.3.1 Mass transfer controlled process of sublimation heating and diffusing from two sides

1. Analyses of mass transfer

Take a thin slice of material as an example to analyze the drying process heating and diffusing in both sides. Here, heat transfer inward is from two sides and the mass transfer outward is to two-sides, as shown in Fig. 2.15(d). Simple discussion was made in previous section, whereas only situation inside the drying chamber was discussed. It was supposed that the outside surface temperature of the dried layer T_s remains unchanged and the vapor pressure at the outside surface of the material P_{sw} approximates the water vapor pressure of the drying chamber P_{ch} ; the values of P_{sw} and P_{ch} were constant during the sublimation stage.

In fact, the drying chamber is conneted to a cold trap and a vacuum pump. Suppose that the surface temperature of cold trap is T_a ; the temperature inside the dried layer T_d ; the temperature inside frozen layer T_f ; and the temperature of radiation heating plates T_{up} , as shown in Fig.3.4.

Suppose the heat transfer condition is favorable, but the mass transfer condition is unfavorable. The mass transfer becomes the major drawback, only the mass transfer process is analyzed here.

2. Mass transfer from the sublimation surface to the outside surface of dried layer

As described in the section above, the sublimation rate is

$$G = -\frac{dW}{dt} = b\frac{(P_{iw} - P_{sw})}{x_d} \tag{3.7}$$

where b —the water vapor permeability through the dried layer, $kg/(m \cdot Pa \cdot h)$.

The sublimation rate may also be expressed as

$$G = -\frac{dW}{dt} = \frac{D}{RT} \times \frac{(P_{iw} - P_{sw})}{x_d} \tag{3.21}$$

where D —the diffusion coefficient of vapor inside the dried layer, m^2/s;

T —vapor temperature inside the dried layer, K;

Obviously, there is a relation between D and b:

$$D = b \times RT \tag{3.22}$$

In the equation, R is the gas constant of water vapor.

$$R = 8.31 \times 10^3 \frac{Pa \cdot m^3}{K \times kg \cdot mol} / 18 \frac{kg}{kg \cdot mol} = 461.7 \frac{Pa \cdot m^3}{kg \cdot K}$$

For an example, the water vapor permeability through the dried layer b is

$2.23 \times 10^{-4}\ kg/(m \cdot Pa \cdot h)$, at $T = 263K$

The diffusion coefficient of vapor inside dried layer D is

$$D = b \times RT = 2.23 \times 10^{-4}\ kg/(m \cdot Pa \cdot h) \times 263K \times 461.7 \frac{Pa \cdot m^3}{kg \cdot K}$$

$$= 27.1\ m^2/h = 75 \times 10^{-4} m^2/s$$

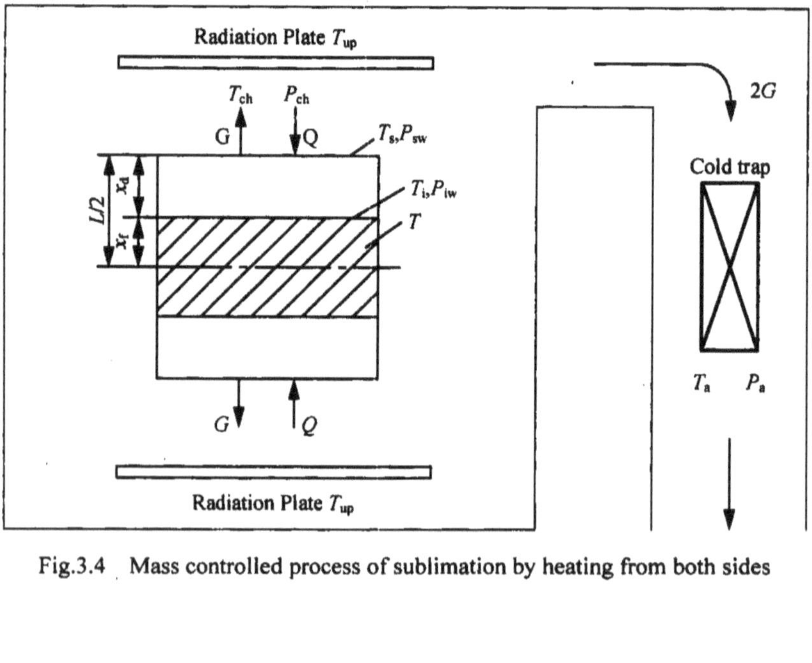

Fig.3.4 Mass controlled process of sublimation by heating from both sides

3. Mass transfer from the outside surface of dried layer to the cold trap

The sublimation rate may also be expressed as

$$G(t) = -\frac{dW}{dt} = \frac{\alpha_m}{RT} \times (P_{sw} - P_a) \tag{3.23}$$

where α_m ——the convective mass transfer coefficient from the outside surface of dried layer to the cold trap, m/s;

T ——the vapor temperature in the zone from the outside surface of dried layer to the cold trap, K.

Combine Eq.(3.21) and Eq.(3.23), there is

$$\frac{D}{RT} \times \frac{(P_{iw} - P_{sw})}{x_d} = \frac{\alpha_m}{RT} \times (P_{sw} - P_a) \tag{3.24}$$

The temperatures of T at the left and right sides of Eq. (3.24) correspond to the temperatures in Eq. (3.21) and Eq. (3.23) respectively. Their values are different. If both take the same value, Eq. (3.24) can be simplified as

$$P_{sw} = \frac{DP_{iw} + \alpha_m x_d P_a}{x_d \alpha_m + D} \tag{3.25}$$

It is obvious from the equation that even if the vapor pressure on the cold trap P_a and the sublimation surface pressure P_{iw} remains unchanged, the vapor pressure on the dried layer surface P_{sw} is still not a constant but a function of dried layer thickness x_d. There are also two parameters, α_m and D, in the function.

4. Drying time and the residual moisture content

Introducing Eq.(3.25) into Eq.(3.23), the sublimation rate becomes

$$G(t) = -\frac{dW}{dt} = \frac{D\alpha_m}{RT} \times \frac{(P_{iw} - P_a)}{x_d \alpha_m + D} \tag{3.26}$$

Let

$$Y = \frac{m - m_e}{m_0 - m_e} \tag{3.27}$$

where Y ——the ratio of the residual moisture to the original moisture at time point t, %;

m ——the residual moisture at time point t, g_{water}/g_{dry};

m_0 ——the moisture before sublimation(t=0), g_{water}/g_{dry};

m_e ——the moisture after sublimation, g_{water}/g_{dry}.

Assuming the moisture distribution inside the material is uniform, the ratio of the moisture already removed to the original moisture content, (1—Y), is proportional to the thickness of dried layer x_d. Namely,

$$\frac{x_d}{L/2} = 1 - Y \tag{3.28}$$

If moisture content is very low after sublimation, i.e., $m_e \to 0$; Eq.(3.27) can be simplified as

$$Y = \frac{m}{m_0} \tag{3.29}$$

and

$$x_d = \frac{L}{2}(1 - \frac{m}{m_0}) \tag{3.30}$$

The sublimation rate G has a relation with the change rate of Y, that is

$$G(t) = -\rho_d m_0 \times \frac{L}{2} \times \left(\frac{dY}{dt}\right) \tag{3.31}$$

where ρ_d——desity of the dried layer, kg_{dry}/m^3;

$L/2$——half thickness of the material, m.

Combining the Eq. (3.26) and (3.31) gives

$$\frac{D\alpha_m}{RT} \times \frac{(P_{iw} - P_a)}{x_d\alpha_m + D} = -\rho_d m_0 \times \frac{L}{2} \times \left(\frac{dY}{dt}\right) \tag{3.32}$$

Substitute x_d of Eq.(3.28) into Eq.(3.32), and then reorganize

$$1 - Y = C_{2m} \times \frac{1}{\left(-\frac{dY}{dt}\right)} - \frac{2D}{\alpha_m L} \tag{3.33}$$

where

$$C_{2m} = \frac{4D(P_{iw} - P_a)}{RTL^2 m_0 \rho_d} \tag{3.34}$$

Rewrite Eq. (3.33) as

$$-C_{2m} dt = (1 - Y + \frac{2D}{\alpha_m L})dY \tag{3.35}$$

Suppose $(P_{iw} - P_a)$ and other parameters in Eq. (3.34) are constants, then C_{2m} is also a constant. Integrate the two sides of Eq. (3.35), the relation of Y and t is obtained as Eq. (3.36) or Eq. (3.37).

$$t = \frac{1}{2C_{2m}}\left((1 - Y)^2 + \frac{4D}{\alpha_m L}(1 - Y)\right) \tag{3.36}$$

or

$$(1-Y)=2C_{2\mathrm{m}}\times\left(\frac{t}{1-Y}\right)-\frac{4D}{\alpha_{\mathrm{m}}L} \tag{3.37}$$

This is the process of mass controlled sublimation by heating and diffusion from both sides

3.3.2 Mass transfer controlled process of sublimation heating and diffusing from one side

Analyses of mass transfer

Some cases of sublimation heating and diffusing from one side are shown in Fig.2.15(a) and Fig.2.15(b). In Fig.2.15(a), a conductive heating is applied to the frozen layer; while in Fig.2.15(b), a radiation heating is applied to the dried layer. The heat transfer models of these two cases are different, however their mass transfer models are same and they are all through the dried layer. Therefore for the mass transfer controlled sublimation process, the mathematical models in these two cases are the same.

Eq.(3.28) and (3.31) are used for heating and diffusing from two sides. It is necessary to make some alterations as below to illustrate the case where heating and diffusing are from one side.

Let

$$x_{\mathrm{d}}=(1-Y)L \tag{3.38}$$

$$G=-\rho_{\mathrm{d}}m_0\times L\times\left(\frac{\mathrm{d}Y}{\mathrm{d}t}\right) \tag{3.39}$$

Using a similar method as in the previous section gives

$$\frac{D\alpha_{\mathrm{m}}}{RT}\times\frac{(P_{\mathrm{iw}}-P_{\mathrm{a}})}{x_{\mathrm{d}}\alpha_{\mathrm{m}}+D}=-\rho_{\mathrm{d}}m_0\times L\times\left(\frac{\mathrm{d}Y}{\mathrm{d}t}\right) \tag{3.40}$$

Substitute x_{d} of Eq. (3.38) into Eq. (3.40), and then reorganize

$$1-Y=C_{1\mathrm{m}}\times\frac{1}{\left(-\frac{\mathrm{d}Y}{\mathrm{d}t}\right)}-\frac{2D}{\alpha_{\mathrm{m}}L} \tag{3.41}$$

where

$$C_{1\mathrm{m}}=\frac{D(P_{\mathrm{iw}}-P_{\mathrm{a}})}{RTL^2m_0\rho_{\mathrm{d}}} \tag{3.42}$$

Rewrite Eq. (3.41) as

$$-C_{1\mathrm{m}}\mathrm{d}t=\left\{(1-Y)+\frac{D}{\alpha_{\mathrm{m}}L}\right\}\mathrm{d}Y \tag{3.43}$$

Suppose $(P_{iw} - P_a)$ and other parameters in Eq. (3.42) are constants, hence C_{lm} is also a constant. From Eq.(3.43), the integral of t and Y respectively ($t:0 \to t; Y:1 \to Y$.) can be found as Eq. (3.44) or Eq. (3.45).

$$t = \frac{1}{C_{lm}}\left(\frac{1}{2}(1-Y)^2 + \frac{D}{\alpha_m L}(1-Y)\right) \tag{3.44}$$

or

$$(1-Y) = 2C_{lm} \times \left(\frac{t}{1-Y}\right) - \frac{2D}{\alpha_m L} \tag{3.45}$$

This is the equation of mass transfer controlled sublimation process with heating and diffusion from one side.

3.3.3 Determination of the parameters α_m and D by experiment[58]

There are two parameters, D and α_m, in the Eq.(3.36),(3.37),(3.44)and(3.45), where D is the vapor diffusion coefficient inside the dried layer; and α_m is the convective mass transfer coefficient from the outside surface of dried layer to the cold trap. These two parameters can not be determined by calculation. They are usually determined from experiments by measuring material weight loss and data organization according to the above sublimation equations.

For example, a frozen material is put on a heating plate to be sublimated. The material has initial moisture content w_0 of 50% and thickness of 1.55cm. The temperature of frozen layer is 250.2K; the density of dried layer is $\rho_d = 800\text{kg/m}^3$. Temperature of the heating plate is 303K and the surface temperature of cold trap is 213.3K.

In the sublimation experiment, the mass variation of material with time is shown in Table 3.4 and Fig. 3.5. Some computational results, such as $(1-Y)$ and $\left(\frac{t}{1-Y}\right)$, are also listed in the table.

Table 3.4 variation of a material mass with time during drying

drying time t(h)	mass of the material M (kg)	mass of moisture M_{water} (kg_{water})	dry basis moisture content m (kg_{water}/kg_{dry})	Y	$1-Y$ in Eqs.(3.37) and (3.45)	$\frac{t}{1-Y}$ in Eqs.(3.37) and (3.45)
0	5.13	2.565	1.000	1.000	0.000	0.000
4	4.80	2.235	0.871	0.871	0.129	31.1
8	4.58	2.015	0.786	0.786	0.214	37.3
12	4.36	1.795	0.700	0.700	0.300	40.0

continued

drying time t(h)	mass of the material M (kg)	mass of moisture M_{water} (kg_{water})	dry basis moisture content m (kg_{water}/kg_{dry})	Y	$1-Y$ in Eqs.(3.37) and (3.45)	$\frac{t}{1-Y}$ in Eqs.(3.37) and (3.45)
16	4.16	1.595	0.622	0.622	0.378	42.3
20	3.96	1.395	0.544	0.544	0.456	43.8
24	3.75	1.185	0.462	0.462	0.538	44.6
28	3.55	0.985	0.384	0.384	0.616	45.5
32	3.35	0.785	0.306	0.306	0.694	46.1
36	3.15	0.585	0.228	0.228	0.772	46.6
40	2.97	0.405	0.158	0.158	0.842	47.5
44	2.82	0.255	0.099	0.099	0.901	48.9
48	2.72	0.155	0.060	0.060	0.940	51.1
52	2.59	0.025	0.010	0.010	0.990	52.5
56	2.58	0.015	0.006	0.006	0.994	56.3
60	2.58	0.015	0.006	0.006	0.994	60.4

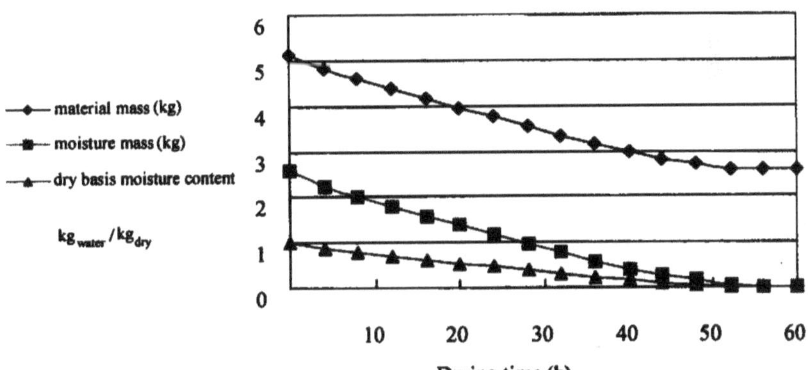

Fig.3.5 Variation of a material mass with time during drying

These are the results of a sublimation drying heating from one side. In the process, heat is supplied on the frozen layer, water vapor escapes through the dry layer, and mass transfer is much more difficult. This process can be analyzed by the mass transfer controlled model as shown in Eq. (3.45).

The data of $(1-Y)$ and $\left(\frac{t}{1-Y}\right)$ calculated from the said mass loss experiment are shown in Table 3.4 and Fig.3.6.

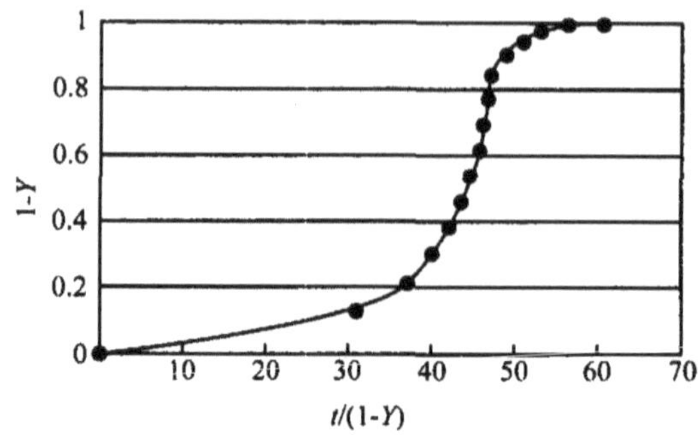

Fig.3.6 Relationship of $t/(1-Y)$ and $(1-Y)$ during sublimation process of a material

Take several points (they are the 5th point to the 12th point here) for linear regression, these 8 points are approximately linear in the chart. A regression equation is obtained from the straight line section of the curve

$$(1-Y)=0.0875\frac{t}{(1-Y)}-3.3449 \tag{3.46}$$

Comparing Eq. (3.46) with Eq. (3.45), the following results are given:

$$C_{1m}=\frac{D(P_{iw}-P_a)}{RTL^2\rho_d w_0}=\frac{0.0875}{2}=0.04375 \tag{3.47}$$

The temperatures of frozen layer and cold trap are 250.2K and 213.3K respectively. The corresponding saturated pressures of ice are P_{iw} =77.557 Pa and P_a=1.104Pa respectively from Table 2.3.

Introducing these data into Eq. (3.47) gives

$$D=\frac{0.04375\times461.7\times250.2\times0.0155^2\times800\times1}{77.557-1.104}$$
$$=12.7(\mathrm{m^2/h})=0.0035(\mathrm{m^2/s})$$

Also from the comparation of Eq. (3.46) and Eq. (3.45), there are

$$\frac{2D}{\alpha_m L}=3.3449$$

and
$$\alpha_m=\frac{2\times0.0035}{0.0155\times3.3449}=0.135(\mathrm{m/s})$$

Regarding other materials having different thickness L(m), C_{1m} is no longer the constant as that in Eq. (3.47); it is inversely proportional to L^2. If the technological conditions of sublimation drying is the same as the above, the values of α_m and D remain unchanged. Substitute the values of α_m and D into Eq.(3.44), and let $Y=0$ (corresponding to the end of sublimation drying), the relation of drying time t (h) with material thickness L(m) can be obtained as below

$$t = 47569 \cdot L^2 + 2466 \cdot L \tag{3.48}$$

Results above are based on the linear extrapolation in Fig.3.6. It fits only for the given drying conditions and does not have universal significance.

3.3.4 Heat transfer controlled drying model [59]

Again take a thin slice of material as an example to analyze the drying process as shown in Fig. 2.15(d), in which the heat transfer inward is from two-side surfaces and the mass transfer outward is from two-side surfaces. Fig.3.1 is a sketch to analyze this process.

Let's denote the drying chamber temperature as T_{ch}, its vapor pressure as P_{ch}; the sublimation interface temperature as T_i and its vapor pressure as P_{iw}; the dried layer thickness as x_d and its outside surface temperature as T_s. For this case, the heating flux Q (W/m^2) is

$$Q = \frac{K_d}{x_d}(T_s - T_i) \tag{3.49}$$

where K_d is the thermal conductivity of dried layer of material, $kW/(m \cdot K)$.

Moreover,

$$Q = \alpha(T_{ch} - T_s) \tag{3.50}$$

where α is the heat transfer coefficient between the gas (with temperature T_{ch}) in the drying chamber and the outside surface of dried layer, $W/(m^2 \times K)$.

For heat transfer from the gas in the drying chamber to the sublimation interface, the heating flux can be written as

$$Q = K(T_{ch} - T_i) \tag{3.51}$$

where K ——the heat transfer coefficient between the gas in the drying chamber and the sublimation interface of the material, $W/(m^2 \times K)$.

Combining Eqs.(3.49), (3.50) and (3.51) gives

$$\frac{1}{K} = \frac{1}{\alpha} + \frac{x_d}{K_d} \tag{3.52}$$

For these cases of heating and diffusing from both sides, there is an equation like Eq. (3.27)

$$\frac{x_d}{L/2}=1-Y$$

where Y is the ratio of the residual moisture to the original moisture content at time point t, %.

Therefore,

$$K=\frac{K_d}{K_d/\alpha+L(1-Y)/2} \tag{3.53}$$

According to heat balance and Eq. (3.31), we get

$$Q=K(T_{ch}-T_i)=G\times\Delta H_s=\Delta H_s\times\rho_d m_0\times\frac{L}{2}\times(-\frac{dY}{dt}) \tag{3.54}$$

where ΔH_s ——the latent heat of sublimation, kJ/kg_{water}.

Introducing Eq. (3.53) into Eq. (3.54) gives

$$\frac{K_d}{K_d/\alpha+L(1-Y)/2}\times(T_{ch}-T_i)=\Delta H_s\times\rho_d m_0\times\frac{L}{2}\times(-\frac{dY}{dt}) \tag{3.55}$$

Let

$$C_{2h}=\frac{4K_d(T_{ch}-T_i)}{\Delta H_s L^2 m_0\rho_d} \tag{3.56}$$

Eq. (3.55) can be rewritten as

$$(1-Y)=\frac{C_{2h}}{(-dY/dt)}-\frac{2K_d}{\alpha\times L} \tag{3.57}$$

From Eq. (3.57) integral of t and Y ($t:\cdot 0\rightarrow t;\cdot and\cdot Y:1\rightarrow Y$;), it is obtained that

$$t=\frac{1}{2C_{2h}}\left((1-Y)^2+\frac{4K_d}{\alpha L}(1-Y)\right) \tag{3.58}$$

or

$$(1-Y)=2C_{2h}\times(\frac{t}{1-Y})-\frac{4K_d}{\alpha L} \tag{3.59}$$

This is the equation of heat transfer controlled sublimation by heating and diffusion from both sides.

For example, a thin slice of frozen material is hung inside the drying chamber to carry on sublimation by heating and diffusion from both sides. Suppose the material has an initial moisture content w_0 = 75%, initial density $\rho_0=965kg/m^3$; the frozen material has a thickness L= 2.50 cm, temperature −25℃; the temperature inside drying chamber $T_{ch}=27$ ℃; the thermal conductivity of the dried layer

$$K_d=0.0692W/(m\cdot K)=6.92\times10^{-5}kW/(m\cdot K)$$

Estimate the drying time when the final moisture content $w_e = 4\%$.

It is assumed that the heat transfer resistance between the drying chamber gas and the outside surface of dried layer is approximately equivalent to the thermal conductive resistance of 3 mm thickness vapor layer. As the thermal conductivity of water vapor is 0.0235 kW/(m·K), the heat transfer coefficient α between the drying chamber gas and the outside surface of dried layer is found as

$$\alpha = \frac{0.0235}{0.003} = 7.833\,\text{W}/(\text{m}^2\cdot\text{K}) = 7.83\times10^{-3}\,\text{kW}/(\text{m}^2\cdot\text{K})$$

Since the initial moisture content w_0 = 75% and the final moisture content $w_e = 4\%$, the initial dry basis moisture content $m_0 = \frac{w_0}{1-w_0} = 0.75/(1-0.75) = 3.0$; after drying time t, the final dry basis moisture content $m_e = \frac{w_e}{1-w_e} = 0.04/(1-0.04) = 0.0417$.

The ratio of the residual moisture content to the original moisture content is

$$Y = \frac{m_e}{m_0} = \frac{0.0417}{3.0} = 0.0139$$

The density of the dried layer is

$$\rho_d = \rho_0/(1+w_0) = 965/(1+3) = 241.25(\text{kg}/\text{m}^3)$$

And the latent heat of sublimation of ice at −25℃ is 2838.4kJ/kg。

Introducing the known values into Eq. (3.56) gives

$$C_{2h} = \frac{4K_d(T_{ch}-T_i)}{\Delta H_s L^2 m_0 \rho_d} = \frac{4\times6.92\times10^{-5}\times(27+25)}{2838.4\times(0.025)^2\times3.0\times241.25} = 1.12\times10^{-5}\ (1/\text{s})$$

and

$$\frac{4K_d}{\alpha L} = \frac{4\times6.92\times10^{-5}}{7.83\times10^{-3}\times0.025} = 1.414$$

Introducing these values into Eq.(3.58) gives

$$t = \frac{1}{2C_{2h}}\left[(1-0.0139)^2 + 1.414\times(1-0.0139)\right]$$

$$= \frac{1}{2\times1.12\times10^{-5}}\left[(1-0.0139)^2 + 1.414\times(1-0.0139)\right]$$

$$= 1.05\times10^5\,\text{s} \cong 29.2(\text{h})$$

3.4 Mathematical model and computational analyses of primary and secondary drying stages

For sublimation drying (the primary drying), heating is supplied under vacuum and low temperature. The heat required in primary drying is mainly for sublimation of frozen ice, but sometimes it is also possible that a small amount of bound water (in unfrozen state) evaporates. Some research indicated that the amount of bound water evaporated in the primary drying stage is very small and its effect on the sublimation drying can be ignored.

For desorption drying (the secondary drying), heating is supplied under vacuum and elevated temperature. Heat required in secondary drying is mainly used for desorption and evaporation of the bound water. Sometimes there is also possibly a small amount of frozen water left after primary drying, that will melts and evaporats at elevated temperature during secondary drying.

This shows that the mechanism and operating parameters in the primary and the secondary drying stages are entirely different. Therefore, to decide when primary drying should end and when the second-drying should start is very important. Too early or too late "switch" will result in the degradation of freeze-dried quality and more consumption of energy and time.

Usually there are two ways to load freeze-dried materials, namely, disk loading (in trays) and bottle loading (in vials). Here the analysis is only for loading in trays. For loading in bottles or in vials, the mathematical model and the essential analysis method are similar to that of disk loading. The Readers who are interested in bottle loading may refer to the related literature[60].

3.4.1 Mathematical model and analysis of primary drying

The material is placed in a tray and the tray is put on a lower plate. The tray is in direct contact with the plate. Heat is supplied in two ways: one is conductive heat transfer from the lower plate through the tray to the material; the other is radiation heat transfer from an upper plate to the material, as shown in Fig.3.7.

For most cases, the diameter of material is much larger than the height of material, therefore the lateral heat transfer may be neglected, and the heat and mass transfer problem can be simplified as a one-dimensional problem.

Literature[61] analyzed the drying of whole milk (skin milk) placed in a tray. The diameter of material D=15cm; the height of frozen material L = 2cm.

During sublimation drying, it is supposed that the lower conductive heating plate has a temperature T_{lp} = 40 ℃; the upper radiation heating plate T_{up} =40℃; the melting temperature and the scorch temperature of whole milk are T_{m}=−10℃ and T_{scor} =60℃, respectively; the effective heat capacity of the dried layer C_{pde}= 2.59 kJ/(kg·K); the heat capacity of frozen layer C_{pf} = 1.93 kJ/(kg·K); the density of frozen layer ρ_f =1030.0 kg/m^3; the density of dried layer ρ_{d} =328.0 kg/m^3 during primary drying and it decreases from 328.0 kg/m^3 to 215.0 kg/m^3 during secondary drying. The initial bound water content in dried layer m_{s}^0 = 0.6415 $\mathrm{kg_{water}/kg_{dry}}$. The sublimation heat of ice ΔH_{s}=2840.0 kJ/kg and the desorption heat of bound water ΔH_{v} = 2687.4 kJ/kg.

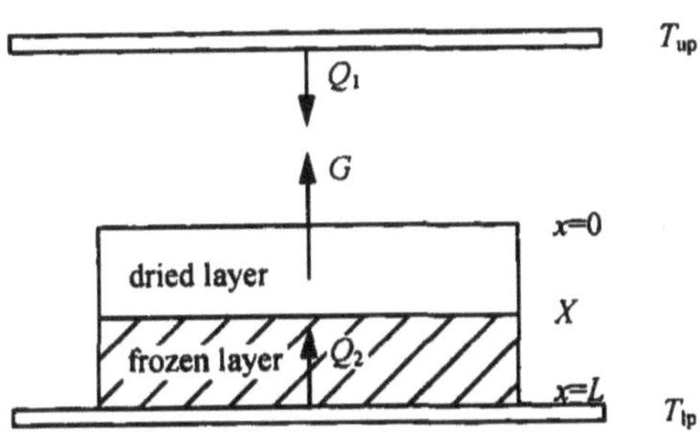

Fig.3.7 The analysis sketch for primary drying of plate material heated by both conduction and radiation from two sides

Regarding more general analyses, there is small amount of bound water (m_{sw}) desorbed during primary drying. In addition to the water vapor flow rate G_{w}, there are also some non-condensable gases extracted from the material at a flow rate G_{in}.

Therefore, the total flow rate G_{t} $\left[\mathrm{kg/(m^2\cdot s)}\right]$ is

$$G_{\mathrm{t}} = G_{\mathrm{w}} + G_{\mathrm{in}} \tag{3.60}$$

1. Energy equation inside dried layer

Suppose the properties of material are uniform and the sublimation does not exist inside the dried layer.

$$\frac{\partial T_d}{\partial t}=a_{de}\frac{\partial^2 T_d}{\partial x^2}-\frac{C_{pg}}{\rho_{de}C_{Pde}}\frac{\partial(G_t T_d)}{\partial x}+\frac{\Delta H_V \rho_d}{\rho_{de}C_{Pde}}\frac{\partial m_{sw}}{\partial t}\quad t>0,0\leqslant x\leqslant X \quad (3.61a)$$

(In addition to the symbols already explained, there are some new symbols as follows)

where $a_{de}=k_{de}/\rho_{de}C_{Pde}$ —— the effective thermal diffusivity of dried layer, m^2/s;

k_{de} ——effective thermal conductivity of dried layer, $kW/(m\cdot K)$;

C_{pg} ——heat capacity of the gases insde dried layer, $kJ/(kg\cdot K)$.

The last term at the right side of Eq. (3.61a) refers to the desorption phenomenon which occurs in primary drying. Some researches have found that it has extremely small impact on the calculation and can be ignored. Therefore Eq.(3.61a) can be simplified as

$$\frac{\partial T_d}{\partial t}=a_{de}\frac{\partial^2 T_d}{\partial x^2}-\frac{C_{pg}}{\rho_{de}C_{Pde}}\frac{\partial(G_t T_d)}{\partial x}\quad t>0,0\leqslant x\leqslant X \quad (3.61b)$$

2. Mass equation inside dried layer

1) Mass equation of water vapor

$$\frac{1}{R}\frac{\partial}{\partial t}\left(\frac{P_{dw}}{T_d}\right)=-\frac{1}{M_w\varepsilon}\frac{\partial G_W}{\partial x}-\frac{\rho_d}{M_W\varepsilon}\frac{\partial m_{sw}}{\partial t}\quad t>0,\ 0\leqslant x\leqslant X \quad (3.62a)$$

where R ——the molar gas constant, $R=8.314$ $kJ/(kmol\cdot K)$;

M_w ——the molecular weight of water;

P_{dW} ——the partial pressure of water vapor inside the dried layer;

ε —— the voidage fraction of dried layer.

The last term at the right side of Eq.(3.62a) refers to the desorption phenomenon which occurs in primary drying. Some researches have also found that it has extremely small impact to the calculation and can be ignored. Eq.(3.62a) can be simplified as

$$\frac{1}{R}\frac{\partial}{\partial t}\left(\frac{P_w}{T_d}\right)=-\frac{1}{M_w\varepsilon}\frac{\partial NG_w}{\partial x}\quad t>0,\ 0\leqslant x\leqslant X \quad (3.62b)$$

2) Mass equation of non-condensable gas

$$\frac{1}{R}\frac{\partial}{\partial t}\left(\frac{P_{in}}{T_d}\right)=-\frac{1}{M_{in}\varepsilon}\frac{\partial G_{in}}{\partial x}\quad t>0,\ 0\leqslant x\leqslant X \quad (3.63)$$

where M_{in} ——molecular weight of non-condensable gas. For air, $M_{in}=29$ $kg/kmol$.

3. Energy equation inside frozen layer

$$\frac{\partial T_\mathrm{f}}{\partial t} = a_\mathrm{f} \frac{\partial^2 T_\mathrm{f}}{\partial x^2} \quad t>0,\ X \leqslant x \leqslant L \tag{3.64}$$

where a_f ——the thermal diffusivity of frozen layer, $\mathrm{m^2/s}$.

There is no mass transfer inside frozen layer.

4. About G_w and G_in inside the dried layer

G_w and G_in occur in the mass Eq. of (3.62b)and(3.63)respectively. Obviously, they are closely related to the structure of the dried layer.

Using the dusty−gas model equations, G_w and G_in are expressed as the following[62]

$$G_\mathrm{w} = -\frac{M_\mathrm{w}}{RT_\mathrm{d}}\left[K_1 \frac{\partial P_\mathrm{w}}{\partial x} + K_2 P_\mathrm{w}\left(\frac{\partial P_\mathrm{w}}{\partial x} + \frac{\partial P_\mathrm{in}}{\partial x}\right)\right] \tag{3.65}$$

$$G_\mathrm{in} = -\frac{M_\mathrm{in}}{RT_d}\left[K_3 \frac{\partial P_\mathrm{in}}{\partial x} + K_4 P_\mathrm{in}\left(\frac{\partial P_\mathrm{w}}{\partial x} + \frac{\partial P_\mathrm{in}}{\partial x}\right)\right] \tag{3.66}$$

The complex influences of dried layer structure to the flow rates can be reduced to four coefficients : K_1, K_2, K_3, K_4 [63]. provided the equations to calculate these four coefficients.

Introducing the Eqs. (3.65) and (3.66) for G_w and G_in into the Eq. (3.62b) and (3.63) respectively, the partial differential equations of P_w and P_in are obtained as follows.

$$\varepsilon \frac{\partial}{\partial t}\left(\frac{P_\mathrm{w}}{T_\mathrm{d}}\right) = \frac{\partial}{\partial x}\left\{\frac{1}{T_\mathrm{d}}\left[K_1 \frac{\partial P_\mathrm{w}}{\partial x} + K_2 P_\mathrm{w}\left(\frac{\partial P_\mathrm{w}}{\partial x} + \frac{\partial P_\mathrm{in}}{\partial x}\right)\right]\right\} \quad t>0,\ 0 \leqslant x \leqslant X \tag{3.67}$$

$$\varepsilon \frac{\partial}{\partial t}\left(\frac{P_\mathrm{in}}{T_\mathrm{d}}\right) = \frac{\partial}{\partial x}\left\{\frac{1}{T_\mathrm{d}}\left[K_3 \frac{\partial P_\mathrm{in}}{\partial x} + K_4 P_\mathrm{in}\left(\frac{\partial P_\mathrm{w}}{\partial x} + \frac{\partial P_\mathrm{in}}{\partial x}\right)\right]\right\} \quad t>0,\ 0 \leqslant x \leqslant X \tag{3.68}$$

5. Initial conditions of primary drying (t=0, 0≤ x ≤ L)

$$T_\mathrm{x} = T^0 \tag{3.69a}$$

$$P_\mathrm{w} = P_\mathrm{w}^0,\ P_\mathrm{in} = P_\mathrm{in}^0,\ P^0 = P_\mathrm{w}^0 + P_\mathrm{in}^0 \tag{3.69b}$$

$$m_\mathrm{sw} = m_\mathrm{s}^0 \tag{3.69c}$$

where P_w^0 —— mainly determined by the temperature and capacity of the cold trap of drying system;

P^0 ——determined by the capacity of vacuum system and other factors.

6. Boundary conditions of the primary drying

1) the heat flux from the lower heating plate Q_2

(1) For a given heat flux Q_2, the lower boundary condition of material is

$$Q_2 = k_f \frac{\partial T_f}{\partial x}\Big|_{x=L}, \quad t>0 \tag{3.70a}$$

where k_f ——thermal conductivity of frozen layer.

The thermal conductivity of frozen skin milk can be expressed as a function of temperature [62]

$$k_f = (0.48819/T_f) + 0.4685\times10^{-3} \tag{3.71a}$$

T_f is the absolute temperature of milk, K.

For T_f=−40℃

$$k_f = \frac{0.48819}{233.15} + 0.4685\times10^{-3} = 0.0021 + 0.0004685 \approx 0.00256 \quad [\mathrm{kW/(m\cdot K)}]$$

[13] give the thermal conductivity of frozen skin milk 0.00266 $[\mathrm{kW/(m\cdot K)}]$ at −40℃.

(2) For a fixed lower plate temperature T_{lp}, the lower boundary condition of the equation is

$$Q_2 = \alpha(T_{lP} - T_f)\cdot\Big|_{x=L}, \quad t>0 \tag{3.70b}$$

where α——the heat transfer coeffcient from the lower plate to the material bottom, $\left[\mathrm{kw/(m^2\cdot K)}\right]$.

The value of α can be approximately calculated by the following equation

$$\alpha = 1.5358\times10^{-3}\,\mathrm{P} \tag{3.71b}$$

where P——the pressure of the drying chamber, $\mathrm{N/m^2}$.

2) If there is a upper heating plate with temperature T_{up}, *the heat flux* Q_1 *(kW/m²) is*

$$Q_1 = \sigma F(T_{up}^4 - T_{d(t,0)}^4) \tag{3.72}$$

where $\sigma = 5.67\times10^{-11}$ $\mathrm{kW/m^2\times K^4}$, the radiation constant of black body;

F——the shape factor;

$T_{d(t,0)}$——the absolute temperature of the material upper surface, K.

7. The balance conditions at the sublimation interface($x = X$)

During sublimation drying stage, the sublimation interface is moving, from $x=0$ ($t=0$) to $x=L$ ($t=t_{x=L}$). The energy balance condition at the sublimation interface($x=X$) is

$$K_{\rm f}\frac{\partial T_{\rm f}}{\partial x}\Big|_{x=X}-K_{\rm de}\frac{\partial T_{\rm d}}{\partial x}\Big|_{x=X}+V\left\{\rho_{\rm f}C_{\rm pf}T_{\rm f}\big|_{x=X}-\rho_{\rm d}C_{\rm pd}T_{\rm d}\big|_{x=X}\right\}+G_{\rm in}C_{\rm pg}T_{\rm x}=\Delta H_{\rm S}G_{\rm w} \tag{3.73}$$

where V ——the moving rate of sublimation interface, m/s;

$\Delta H_{\rm S}$ ——the heat of sublimation.

The temperature condition of the sublimation interface is

$$T_{\rm d}=T_{\rm x}=T_{\rm f}\ (\mathrm{t}>0,\ x=X) \tag{3.74}$$

The pressure conditions at the sublimation interface are

$$\frac{\partial P_{\rm in}}{\partial x}\Big|_{x=X}=0 \tag{3.75}$$

$$P_{\rm w}=fun(T_{\rm x}) \tag{3.76}$$

Eq. (3.76) is the property of water as already shown in Table 2.2.

The moving rate of sublimation interface is

$$V=\frac{\mathrm{d}X}{\mathrm{d}t}=\frac{-G_{\rm w}\big|_{x=X}}{\rho_{\rm f}-\rho_{\rm d}} \tag{3.77}$$

Its initial condition is $X=0$ at $t=0$.

8. Numerical solution and its comparison with the test

The above equations can be solved by numerical solution method. [61] made a computation on primary drying of whole milk. The frozen whole milk diameter R=15cm, the height of material L = 2cm. The lower conductive plate has a temperature $T_{\rm lp}$= 40℃; the upper radiation heating plate $T_{\rm up}$ = 40℃.The initial conditions are: $T^0(x)=-40$℃, $P_{\rm w}=P_{\rm w}^0=1.07\ \mathrm{N/m^2}$, $P^0=P_{\rm w}^0+P_{\rm in}^0=5.07\ \mathrm{N/m^2}$, $m_{\rm s}^0=0.6415\ \mathrm{kg_{water}/kg_{dry}}$. The initial voidage fraction of the dried layer $\varepsilon=0.785$. Other conditions and data are the same as mentioned above.

During primary drying the sublimation interface starts to move from the upper surface of the material ($X=0$), and falls down gradually to the bottom ($X=L$). Fig. 3.8 gives the computation result of sublimation drying under above-mentioned conditions, and its comparison with the experimental data; they are well consistent. Since the computation had not taken the bound water desorption into consideration, it is thought that the bound water desorption can be neglected during primary drying.

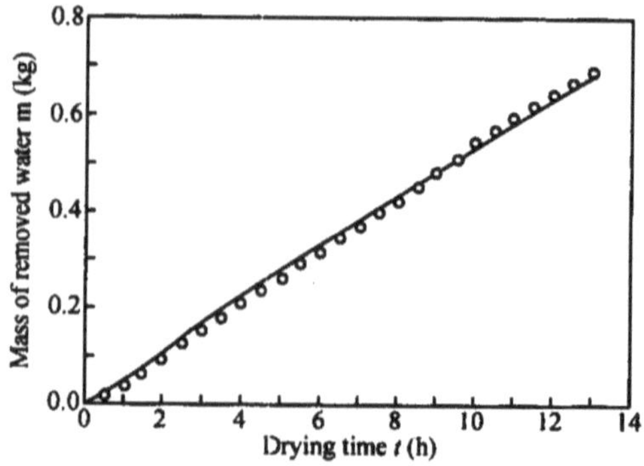

Fig.3.8 The sublimation drying process of a frozen whole milk

the solid line is computation solution; the circles are experimental data

3.4.2 Mathematical model and analysis of secondary drying

After primary drying, the material is heated at elevated temperature to enter the secondary drying. At this stage there is no longer frozen layer, hence the entire material can be considered as a whole.

1. Energy equations of secondary drying

Similar to Eq. (3.61), we have

$$\frac{\partial T_d}{\partial t} = \alpha_{de}\frac{\partial^2 T_d}{\partial x^2} - \frac{C_{pg}}{\rho_{de}C_{Pde}}\frac{\partial (G_t T_d)}{\partial x} + \frac{\Delta H_V \rho_d}{\rho_{de}C_{Pde}}\frac{\partial m_{sw}}{\partial t} \tag{3.78}$$

$$t > t_{X=L}, 0 \leqslant x \leqslant X$$

During secondary drying, the last term at the right side of Eq. (3.78) cannot be ignored; it is the most important one.

The variation of mass fraction of bound water in the dried layer m_{sw} (kg_{water}/kg_{dry}) can be expressed in one of the following two forms

$$\frac{\partial m_{sw}}{\partial t} = f_s\left(m_{sw}^* - m_{sw}\right) \tag{3.79a}$$

or

$$\frac{\partial m_{sw}}{\partial t} = -f_d m_{sw} \tag{3.79b}$$

where m_{sw}^* ——the mass fraction of bound water in the dried layer in equilibrium. [61] gives an approximate equation of m_{sw}^*

$$m_{sw}^* = \exp\{2.3[1.36-0.036(T-T^0)]\}/100 \ (kg_{water}/kg_{dry})$$

where T^0 is the initial temperature; for example it is −40℃ for the case here.

The coefficients f_s , f_d in Eq. (3.79a) and (3.79b)can be obtained by experiments.

2. Mass equations of secondary drying

Similar to Eq. (3.67) and (3.68), we have

$$\varepsilon\frac{\partial}{\partial t}\left(\frac{P_w}{T_I}\right)=\frac{\partial}{\partial x}\left\{\frac{1}{T_d}\left[K_1\frac{\partial P_w}{\partial x}+K_2P_w\left(\frac{\partial P_w}{\partial x}+\frac{\partial P_{in}}{\partial x}\right)\right]\right\}, t>t_{X=L}, 0\leqslant x\leqslant L \quad (3.80)$$

$$\varepsilon\frac{\partial}{\partial t}\left(\frac{P_{in}}{T_d}\right)=\frac{\partial}{\partial x}\left\{\frac{1}{T_d}\left[K_3\frac{\partial P_{in}}{\partial x}+K_4P_{in}\left(\frac{\partial P_w}{\partial x}+\frac{\partial P_{in}}{\partial x}\right)\right]\right\}, t>t_{X=L}, 0\leqslant x\leqslant L \quad (3.81)$$

3. Initial conditions of the secondary drying($t=t_{x=L}$, $0\leqslant x\leqslant L$)

By calculation, the parameters at the end of primary drying $t=t_{x=l}$ can be obtained, such as the temperature distribution $T_d=r(x)$, the partial pressure of water vapor $P_w=\delta(x)$, the partial pressure of non-condensable gases $P_{in}=\theta(x)$, and the mass fraction of bound water in the dried material $m_{sw}=v(x)$. They are the initial conditions of secondary drying, and they are all functions of position x, as shown in the following equations

$$T_d=r(x) \qquad t=t_{x=L}, 0\leqslant x\leqslant L \quad (3.82a)$$

$$P_w=\delta(x) \qquad t=t_{x=L}, 0\leqslant x\leqslant L \quad (3.82b)$$

$$P_{in}=\theta(x) \qquad t=t_{x=L}, 0\leqslant x\leqslant L \quad (3.82c)$$

$$m_{sw}=v(x) \qquad t=t_{x=L}, 0\leqslant x\leqslant L \quad (3.82d)$$

4. Boundary conditions of secondary drying

At the upper surface of the material

$$Q_1=\sigma F_{1-2}\left[T_{up}^4-T_d^4(t,0)\right] \quad t>t_{X=L}, x=0 \quad (3.83)$$

$$P_w=P_{w0}, \quad P_{in}=P_{in0}=P_0-P_{w0} \quad t>t_{X=L}, x=0 \quad (3.84)$$

At the lower surface of material

$$Q_2=k_{de}\frac{\partial T_d}{\partial x}\Big|_{X=L}, \quad x=L, \ t>t_{X=L} \quad (3.85)$$

During secondary drying, k_{de} can be calculated as flows [61, 62]

$$k_{de}=1.412\times 10^{-6}(P_0+P_x)+2.165\times 10^{-4}\ [\text{kW}/(\text{m}\cdot\text{K})] \quad (3.86)$$

When the lower plate temperature T_{lp} and the lower surface temperature of material $T_{d(t,L)}$ are known, the boundary condition of heat transfer at the lower surface of material is

$$Q_2 = \alpha\left(T_{\mathrm{lp}} - T_{\mathrm{d}(t,L)}\right),\quad t > t_{X=L} \tag{3.87}$$

At the same time, there are also following boundary conditions at the lower surface

$$\frac{\partial P_{\mathrm{w}}}{\partial x}\Big|_{x=L} = 0 \quad x = L,\ t > t_{X=L} \tag{3.88}$$

$$\frac{\partial P_{\mathrm{in}}}{\partial x}\Big|_{x=L} = 0 \quad x = L,\ t > t_{X=L} \tag{3.89}$$

5. Numerical solution and its comparison with the experiment

Regarding secondary drying process of whole milk, numerical solutions and experimental data are shown in Fig.3.9, the conditions are the same as previously described. The solid line indicates the numerical solution based on Eq.(3.79a); the dashed line indicates the numerical solution based on Eq. (3.79b); and the circles indicate the experimental data.

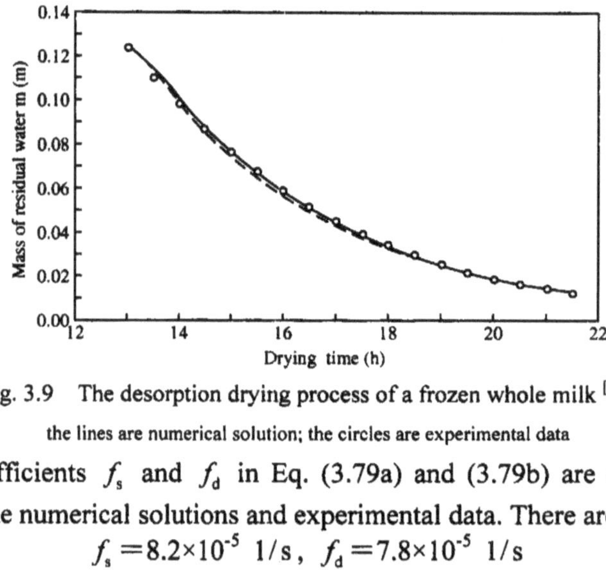

Fig. 3.9 The desorption drying process of a frozen whole milk [61]

the lines are numerical solution; the circles are experimental data

The coefficients f_s and f_d in Eq. (3.79a) and (3.79b) are obtained by fitting of the numerical solutions and experimental data. There are [61]

$$f_s = 8.2\times10^{-5}\ 1/\mathrm{s},\quad f_d = 7.8\times10^{-5}\ 1/\mathrm{s}$$

In above calculation, the results of calculation match well with experimental data without considering convection inside porous material. This indicates that the convection effect in the porous material is negligible.

3.4.3 Analyses of drying process under various operation policies

1. Drying process under various operation policies

Freeze-drying is a time-consuming and energy-consuming process. If drying time can be shortened, energy will be saved. Reasonable variation of operation policies may reduce drying time. The commonly used variable parameters are:

(1) the upper radiation heating plate temperature T_{up} : the heat flux Q_1 varies with T_{up} ;

(2) the lower conduction heating plate temperature T_{lp} : the heat flux Q_2 varies with T_{lp} ;

(3) the drying chamber pressure P_0 . In many cases, heat flux Q_2 is much larger than heat flux Q_1 in Fig.3.7, hence the lower plate temperature T_{lp} is usually adjusted to change the heat flux. The lower plate temperature T_{lp} can be modulated by the power of electric heater or temperature of the liquid inside the lower heating plate.

The pressure of drying chamber P_0 has also great influence on the drying process. Increasing chamber pressure can markedly strengthen heat transfer. However, the interface temperature increases with pressure in the chamber too. Improper control of chamber pressure would harm the quality of freeze dried products[64].

Drying chamber pressure P_0 is mainly composed of two parts: one is the water vapor pressure P_w , the other is the non-condensable gases pressure P_{in} . There is $P_0 = P_w + P_{in}$. Water vapor pressure P_w is mainly determined by the temperature and the ability of cold trap, but also related to the flow resistance of the vacuum system. The lower the condenser temperature, the lower the drying chamber pressure is. The non-condensable gases pressure P_{in} in the drying chamber has two sources: one is air leaking from atmospheric environment; the other is gases escaping from the material itself. A bleeding valve can be installed on the drying chamber and the pressure P_{in} inside the drying chamber can be regulated by adjusting the valve[48, 64].

2. Analyses of primary drying process under various operation policies

In previous analyses, the drying chamber pressure P_0 is maintained at $5.07\ \mathrm{N/m^2}$; the upper radiation plate temperature T_{up} is maintained at 40℃, and the lower conduction plate temperature T_{lp} at 40 ℃. These conditions remain unchanged during primary drying. The results of analyses are: time of primary drying is approximately 13 hours; time of secondary drying is approximately 8 hours; in total the drying takes 21 hours.

[65] analyzed sublimation drying of whole milk on small tray. Its basic parameter were same as those described earlier in this section; however the chamber pressure P_0 is maintained at 1Pa; and the temperature of the lower plate T_{lp} is adjusted in order to regulate the heat flux Q_2 . The parameter variations during primary drying are shown in Figs.3.10 to 3.13.

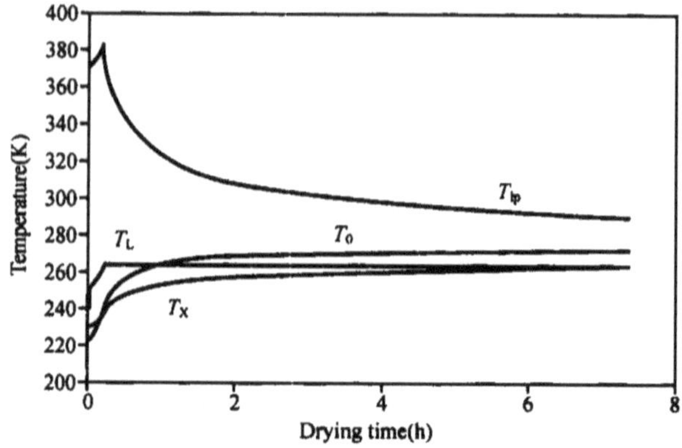

Fig. 3.10 Influence of the plate temperature T_{lp} on temperatures of material during primary drying

T_0, T_X, T_L are the temperatures of the upper surface, the sublimation interface and the lower surface respectively

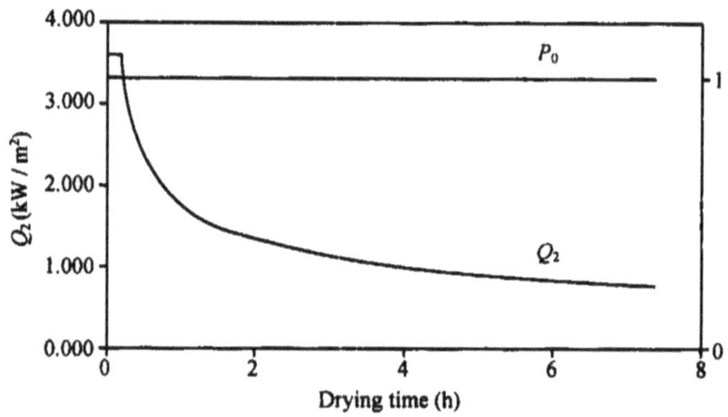

Fig. 3.11 Adjusting the chamber pressure and the heating flux Q_2

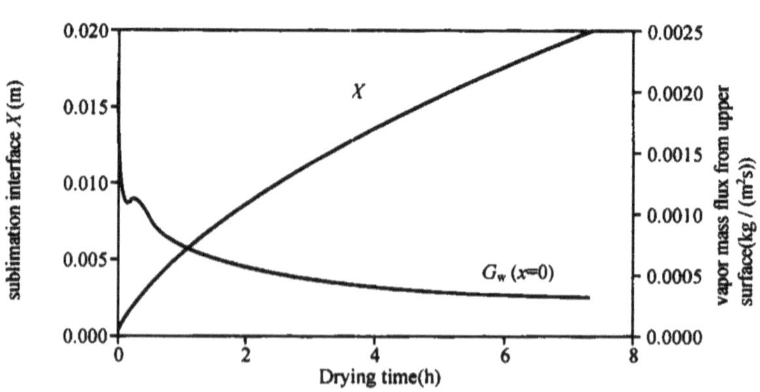

Fig.3.12 Variations of the sublimation interface position and the vapor mass flux

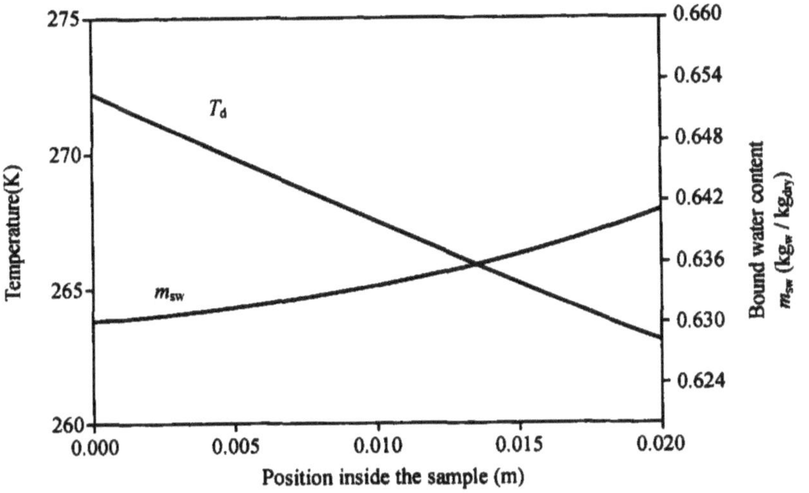

Fig.3.13 Distributions of temperature and moisture at the end of sublimation drying

This analysis concludes that, when plate temperature T_{lp} is lowered, not only the plate power Q_2 may be reduced greatly, but also time for sublimation drying can be reduced from 13h to 7.34h.

3. Analyses of secondary drying process under various operation policies

[65] analyzed also the some operation policies of secondary drying of whole milk, in which the chamber pressure P_0 increased from 1Pa to 1800Pa (the water saturation temperature corresponding to 1800Pa is about 16℃), as shown in Fig. 3.14. Simultaneously, the lower plate temperature T_{lp} was also adjusted as shown in Fig. 3.15.

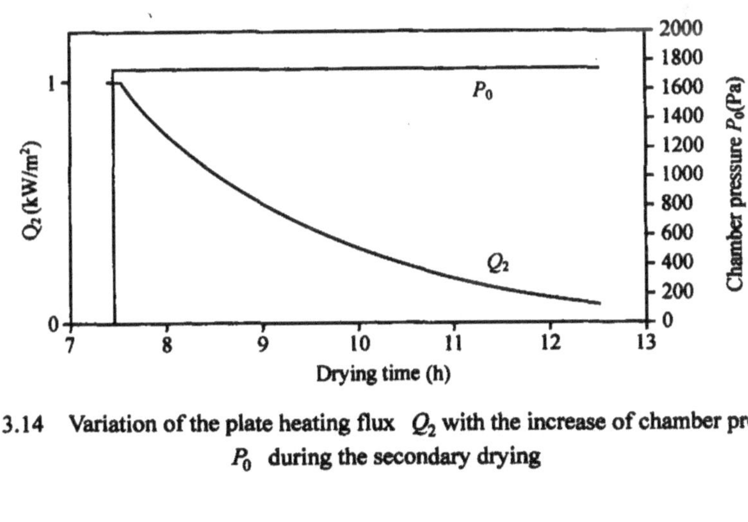

Fig. 3.14 Variation of the plate heating flux Q_2 with the increase of chamber pressure P_0 during the secondary drying

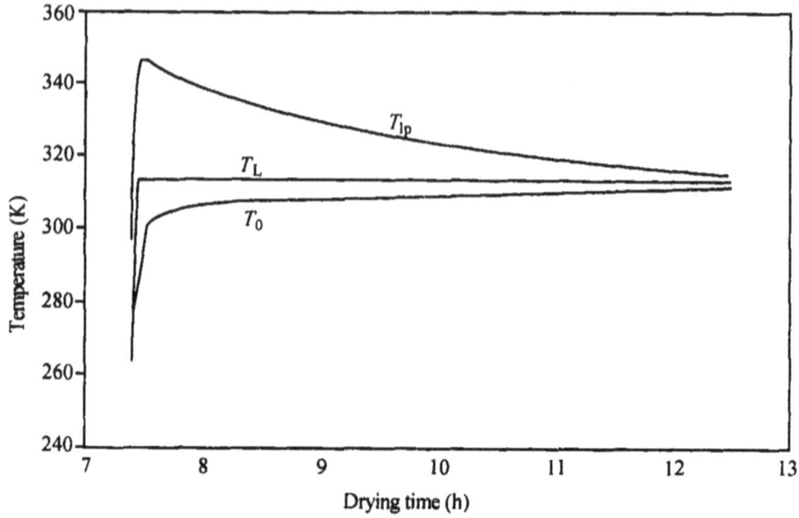

Fig. 3.15 Influence of the lower plate temperature T_{lp} on the temperatures of material during secondary drying

T_0, T_L are the temperatures of upper surface and lower surface, respectively

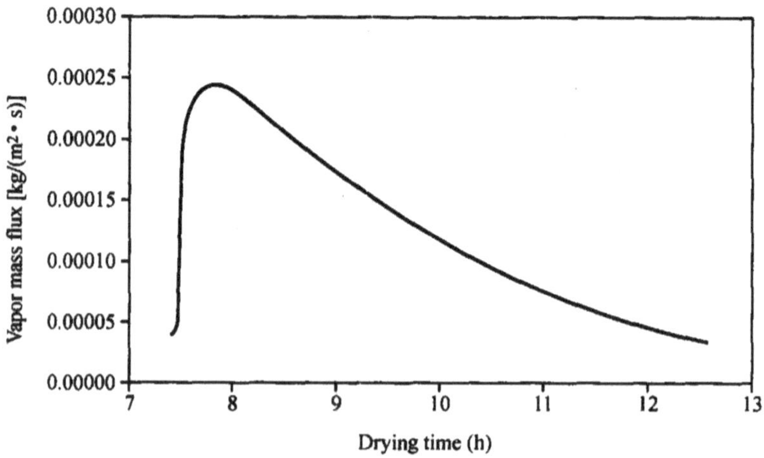

Fig.3.16 Vapor mass flux from the upper surface G_w during secondary drying

Bound water content m_{sw} is reduced approximately from 0.636 kg_{water}/kg_{dry} to 0.049 kg_{water}/kg_{dry} during secondary drying.

This analysis concludes that, when chamber pressure P_0 increases and the plate temperature T_{lp} decreases, not only the plate heating power Q_2 may be reduced greatly, but also the time for secondary drying can be reduced from 8h to 5.2 h. In total time for drying shortens from 21 h to 12.57 h.

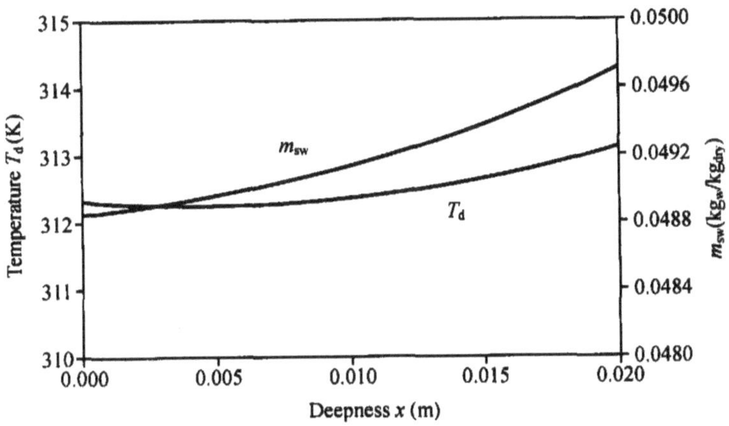

Fig.3.17 Distribution of temperature T_d and moisture content m_{sw} at the end of secondary drying

3.5 Thermal properties of freeze dried material

The thermal properties of freeze dried material are related not only with the components of the material, but also with its structure. Some structure changes are formed during freezing. For example, big ice crystals are formed during slow freezing and large channels are left after sublimation, so the mass transfer resistance is reduced and the drying speeds up. However, when small crystals are formed during rapid freezing and tiny cavities are left after sublimation, the mass transfer resistance increases and the drying speeds down. In addition, during slow freezing, solute migration may occur to such extent that the crust layer form on the surface and it will block the sublimation.

Ideally, water vapor escapes through the cavities and channels in the dried layer which are formed during sublimation. However, if the ice crystals are separated and surrounded by solid matrix, the water vapor can only escape by diffusion or penetration. This situation may also appear when the dried layer collapsed.

The collapse takes place at higher temperatures. When the dried layer is at higher temperature, the viscosity of matrix decreases; the softening and collapse of matrix occur. The collapse temperature is related not only to temperature, but also to the moisture content. The collapse temperature is lower as the moisture content increases.

At the beginning of primary drying, the rate of sublimation is higher since both the heat transfer resistance and the mass transfer resistance are very small. Along with sublimation, the porous dried layer appears and gets thicker. The thermal conductivity of porous dried layer is very low and the effective thermal conductivity of dried layer is related to pressure. Increasing pressure can enhance the effective thermal conductivity, and it is beneficial to heat transfer.

In the drying process, the drying chamber pressure, the heating plate temperature and the cold trap temperature are all important control parameters. A large number of studies indicate that increasing slightly the drying chamber pressure (for example, 266—666 Pa) for a short period of time, and elevating the dry surface temperature close to its allowable value may speed up drying. However the drying chamber pressure must be lower than the sublimation interface pressure and the sublimation temperature must be lower than the vitrification temperature of the material.

1. Effective thermal conductivity of freeze dried material

The effective thermal conductivities K_d of the freeze dried materials are not only related with the solid state matrix structure, but also with the drying chamber pressure and the gas, as shown in Fig. 3.18.

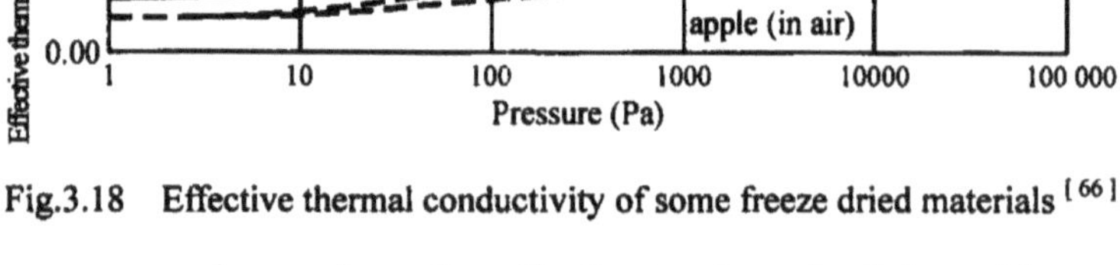

Fig.3.18 Effective thermal conductivity of some freeze dried materials [66]

When pressure is very low, the effective heat conductivity achieves a low asymptotic value which is not relevant to the peripheral gas. Because the gas pressure is so low, the asymptotic value is only related to the solid matrix itself. At a higher pressure, the effective thermal conductivity reaches a quite high asymptotic value, which is not only related to the solid matrix itself, but also to the gas property and pressure. The effective thermal conductivities increase with the thermal conductivity of gas and pressure. When the thermal conductivity achieves high asymptotic value, the mean free path of gas molecular is smaller than the cavity size inside the dried material.

The transition of effective thermal conductivity from low asymptotic value to high asymptotic value is relevant to the ratio of mean free path of the gas molecular to the cavity size of material. The transition usually occurs under 10 Pa—10 kPa. Table 3.5 lists the effective thermal conductivity of some food materials in certain pressure scope and in certain peripheral gas.

Table 3.5 The effective thermal conductivities of freeze dried food materials [56]

material	peripheral gas	Pressure (Pa)	Thermal conductivity [W/(m·K)]
beef	water vapor	67—320	0.035—0.055
mushrooms	air	40—10^5	0.010—0.036
corn starch solution	water vapor and air	13—267	0.014—0.033
beef	Air	0.1—10^5	0.038—0.066
apple	Air	0.1—10^5	0.016—0.042
peach	Air	0.1—10^5	0.016—0.042
peach	Freon-12	3—10^5	0.023—0.187
apple	water vapor	1—40	0.035—0.116
milk	water vapor	1—40	0.023—0.081
salmon	water vapor	20	0.042—0.133
cod	water vapor	10	0.019—0.026
perch	water vapor	10	0.023—0.035
beef	water vapor and air	1—10^4	0.035—0.064
beef	water vapor	27—400	0.052—0.073
potato starch	water vapor and air	4—10^5	0.009—0.042
gelatin	water vapor and air	4—10^5	0.002—0.042
cellulose gum	water vapor and air	4—10^5	0.019—0.055
ovalbumin	water vapor and air	4—101325	0.014—0.042
pectin	water vapor and air	4—101325	0.012—0.050
tomato juice	water vapor	53—200	0.035—0.173

The effective thermal conductivities of freeze dried materials are extremely low as shown in Table 3.5. They are lower than the ordinary thermal insulating materials in many cases. For example, a cork has thermal conductivity of (0.044—0.079) W/(m·K); a foam polystyrene has (0.040—0.043) W/(m·K). Temperature drop across dried layer is quite large because of its very low thermal conductivity. The upper surface temperature is usually controlled under 65℃, sometimes under 38℃. The frozen layer temperature is usually below −18℃ or even lower.

2. Effective diffusivity of water vapor in dried layer

The effective diffusivity D of water vapor in the dried layer is not only related to the matrix structure, but also to the drying chamber pressure and the gas, as shown in Fig.3.19.

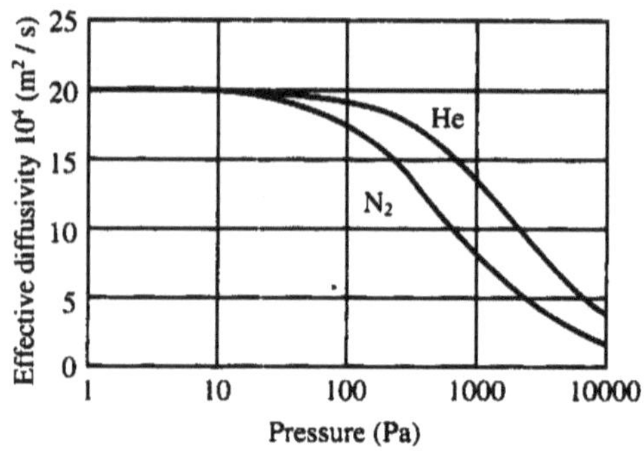

Fig.3.19 The effective diffusivity of freeze dried turkey [67]

Parallel to the direction of muscle fiber

Fig.3.19 indicates that the effective diffusion coefficient D achieves a high asymptotic value under very low pressure; this value corresponds to the Knudsen diffusion, which is relevant the structure of solid matrix. The high asymptotic value is not related to the gas and its pressure.

The Knudsen number is defined as:

$$Kn = \frac{\lambda}{L}$$

where λ——mean free path of the gas, m;

L——representative length of the cavity, m.

Under higher pressure, the effective diffusion coefficient D decreases with the increase of pressure and also related to the property of gas.

3.6 Discussion on mathematical model for drying process

In this chapter, mathematical models for primary drying and secondary drying have been discussed. These models and their numerical calculation have important effect on the qualitative analysis of drying process. However there is quite a gap between the present mathematical models and the practical situation. There are still many problems left in the above models which are briefly discussed in the following.

1) About thermal properties of material

Thermal properties of material are the basis of heat and mass transfer calculation. However thermal properties of the majority of materials are still unknown so far. Regarding the frozen material, their thermal properties may approximate those of ice or approximately calculated by the components of the material[6]. Regarding dried material, their thermal properties data are nearly blank. Thermal properties of dried material are relevant to not only their components, but also the freeze drying process, which has close relation with the structure of the material, such as porosity, etc. These thermal properties of dried material can not be obtained by computation through its components and they are also difficult to be determined through a couple of simple experiments.

2) About temperature measurement during drying process

Measurement of temperature variation and distribution inside small material is extremely difficult. When temperature probes are inserted inside the material, the actual temperature has been changed and the drying process changed too. [4] had made detailed analyses on this issue.

3) About the switch from primary drying to secondary drying

Primary drying is sublimation of frozen water. Temperature of the material in the stage should be quite low (lower than the eutectic temperature or the vitrification temperature of the material). Heating at such low temperature is time-consuming and energy-consuming process. However, in secondary drying the material is heated at elevated temperature for desorption of bound water. The switch time is the moment at which primary drying should be shifted to secondary drying. The determination of the switch time is very important as well as difficult. Too early or too late switch imposes big impact on the quality of the freeze-dried product or wastes a great deal of time and energy. The mathematical models of primary drying and secondary drying should be closely related to the drying stage switch. The method using measurement of dynamic pressure changes to determinate sublimation interface temperature is one of the ways to resolve the problem [4].

4) About pressure and flow rate of non-condensable gases

In the previous analysis, it is supposed that the chamber pressure P is contributed by the water vapor pressure P_w and the noncondensable gas pressure P_{in}, as in Eq. (3.69b), namely $P^0 = P_w^0 + P_{in}^0$. In the Liapis analysis, it is assumed that the pressure of non-condensable gas P_{in} is formed only by the gas escaped from the material and the non-condensable gas mass flux is calculated by this basis[60, 61, 65]. In fact, the non-condensable gas pressure inside the drying chamber is from the gas escaped from material and the gas leakage from atmospheric environment into the vacuum drying chamber.

5) About vitrification in freeze drying process

In the Liapis analysis, the material temperature during primary drying was maintained lower than its melting temperature (for whole milk, the melting temperature T_m=−10℃). The material temperature during secondary drying was maintained lower than its scorch temperature (for whole milk, the scorch temperature T_{sc} =60℃). It was an early theory, however it is no longer suitable, especially for the freeze drying of biological pharmaceuticals and cells. The modern freeze-drying theory is based on the vitrification theory. It asserts that the process of freeze-drying should be in accordance with the way as shown in Fig. 2.10.

In Liapis analyses and experiments, the material is whole milk without any additives. Regarding freeze-drying of modern biological pharmaceuticals, the addition of a variety of components is necessary. Those additives are lyoprotectant, bulking agent, antioxidant, and buffer agent, as shown in Chapter 6 of this book. For those complex materials, the differential scanning calorimetry (DSC) must be used to measure the important parameters of complex material; then the freeze drying process should be carried out according to the vitrification theory. The related mathematical analysis and the computation must also be consistent with such freeze-drying process.

4

Equipment for Freeze-drying

The main components and their technical requirement have been discussed in Chapter 1. In this chapter, the vacuum technique and vapor condenser are firstly introduced, and then several typical freeze-drying devices are presented.

4.1 Vacuum requirement of freeze-drying

4.1.1 Ranges and acquirement of vacuum

A vacuum is a state of a given space in which the gaseous pressure is less than atmospheric pressure. Because the density of gas molecules can not easily be measured, the residual gas pressure is generally the primary indicator of the vacuum quality inside the space.

The unit of pressure in SI system is Pa (Pascal). 1 Pa=1 N/m^2.

In vacuum technology, the practical pressure units also have:

bar and *μ bar*:	1 bar = 10^5 Pa; 1 μ bar = 0.1 Pa
standard atmosphere (atm):	1 atm =1.0133×10^5 Pa
technical atmosphere (at):	1 at = 0.96784 atm = 9.8067×10^4 Pa
torr:	1 torr = 1/760 atm = 1.3332×10^2 Pa

The conversion relations of these pressure units are shown in Table 4.1.

Vacuum quality is usually divided into ranges according to the technology required to achieve it or measure it. These ranges do not have universally agreed definitions, but a typical distribution is:

Rough vacuum or low vacuum: 760 torr to 1 torr (10^5 Pa to 10^2 Pa)

Medium vacuum: 1 torr to 10^{-3} torr (10^2 Pa to 10^{-1} Pa)

High vacuum: 10^{-3} torr to 10^{-7} torr (10^{-1} Pa to 10^{-5} Pa)

Ultra-high vacuum: < 10^{-7} torr (10^{-5} Pa)

The methods of creating vacuum can be divided into the following categories by its principle:

Table 4.1 conversion of pressure units in vacuum technology

pressure units	pascal (Pa)	torr	μ bar	standard atmosphere (atm)	technical atmosphere (at)	pound-force per-square inch (psi)
1 Pa	1	7.5006×10^{-3}	10	9.8692×10^{-6}	1.0197×10^{-5}	145.04×10^{-6}
1 torr	1.3332×10^{2}	1	1.3332×10^{3}	1.3158×10^{-3}	1.3595×10^{-3}	19.337×10^{-3}
1 μ bar	10^{-1}	7.5006×10^{-4}	1	9.8692×10^{-7}	1.0197×10^{-6}	14.5037744×10^{6}
1 atm	1.0133×10^{5}	760	1.0133×10^{6}	1	1.0332	14.696
1 at	9.8067×10^{4}	735.56	9.8067×10^{5}	9.6784×10^{-1}	1	14.223
1 psi	6,894.76	51.715	68.948×10^{3}	68.046×10^{-3}	70.307×10^{-3}	1

By using compression and expansion effects of the gas: mechanical pumps, the vacuum range can be attained to 10^{-2} torr;

By using viscous effects: vapor ejector pumps, 10^{-1} torr to 10^{-5} torr;

By using diffusion effects: vapor diffusion pumps, 10^{-8};

By molecular force: molecular pumps, 10^{-2} to 10^{-10} torr;

By ionization effects, ion pumps, 10^{-4} to 10^{-14} torr;

By cryogenic condensation: cryogenic pumps, 10^{-4} to 10^{-14} torr or higher.

4.1.2 Vacuum ranges in freeze-drying technique

In the Chapter 1 of this book, it has already been discussed that in the drying process of materials large amount of water is sucked away mainly by a water vapor condenser also called "cold trap". The water vapor condenser can provide a low temperature environment for a vacuum system and its temperature is much lower than that of the materials sublimation interface. In this way, the vapor pressure difference between the materials sublimation interface and the vapor condenser surface becomes a driving force of mass-transfer which makes the vapor escape from the materials by sublimation. Meanwhile, the vapor from the materials is condensed into liquid water on the low temperature surface of the condenser and then drained from the system.

In a freeze-drying system, the main function of a vacuum pump is to exhaust gases, which includes the air leaking into the system, the air escaping from the materials, some non-condensation gases. Sometimes vacuum pump also has function to exhaust water vapor which has not been "trapped" by the cold trap.

The requirement of a freeze-drying system to vacuum mainly depends on the temperature of "cold trap". For example, according to Table 2.3 in Chapter 2, if the ice temperature is −30℃, −50℃, −70℃, −100℃, the corresponding saturated vapor pressure will be 38 Pa, 3.9 Pa, 0.26 Pa, 1.4×10^{3} Pa respectively. Currently in the common used freeze dryers the temperatures of water vapor condenser are usually about −40 ℃ or −60 ℃, and the corresponding saturated vapor pressures are 12.85Pa or 1.08Pa. In a freeze-drying system, the commonly used vacuum range is from 1 Pa to 10 Pa, which belongs to the scope of medium vacuum. Therefore, mechanical pumps can be used to obtain the required vacuum.

4.2 Vacuum measurement in freeze-drying

4.2.1 Devices in vacuum measurement

In the vacuum range of freeze-drying, the vacuum measuring devices (also called vacuum gauge) have two main categories: diaphragm manometer and thermal vacuum gauge.

1) Diaphragm manometer

Diaphragm gauge measures pressure differences by the deflection of a flexible metal diaphragm. The pressure of the chamber to be measured makes the metal diaphragm a certain amount of elastic deformation. The deformation can be measured using mechanical, optical or capacitive techniques, however, capacitive technique is most common used at present, which is depends upon the capacitance between the diaphragm and a fixed flat electrode. Movements of the diaphragm, in response to the pressure, change the spacing between diaphragm and electrode, and therefore the capacitance, which can be measured with a capacitance bridge.

The diaphragm manometer measures the total pressure of the space, and it has characteristics of fast dynamic response, high measuring accuracy, good repeatability and stability. The diaphragm manometer can be used in many cases such as air oxygen, water vapor and oil vapor etc.; and the measuring results has no relation with the gas type. The appropriate measuring range of modern capacitance diaphragm manometer is from 0.1 Pa to 1 atm.

2) Thermal vacuum gauge

Thermal vacuum gauges at present mainly refer to the thermal conductivity gauge (Pirani gauge), and also include the convection vacuum gauge.

(1) Thermal conductivity gauge. Thermal conductivity gauge is based on a filament (such as a tungsten or platinum filament, for example, with a diameter of 25 μ m and a length of 57mm) mounted in a glass or metal envelope (slender tube) attached to the vacuum system, the filament being heated by an electric current. The temperature of filament depends on the supply of electrical energy and the heat loss to surrounding. Then the pressure inside the vacuum gauge can be determined by the temperature.

If the slender tube size can reach to millimeter level or less, the natural convection will be prevented (Grashof number $Gr<10^3$); and if the pressure is lower than 10 Pa, so that the gas molecules free path l becomes much larger than the sandwich size L inside the gauge, then the rarefied gas can be out of a state of continuous medium and the gas thermal conductivity is no longer maintain a constant.

The basic principle of thermal conductivity gauge is based on the characteristic that gas thermal conductivity decreases with the pressure decreasing in non-continuous medium state. Therefore thermal conductivity gauge should be applied in non-continuous medium state, especially in free-molecule state (from 10 Pa to 0.01Pa). In principle, this gauge can not used in a too low vacuum or too high vacuum. If the pressure is too high, the gas thermal conductivity will be independent of pressure, and that will cause convection. Whereas if the pressure is too low, the heat loss by radiation and solid conductivity which are independent of pressure will be more than that by gas conductivity, and becomes dominant.

There are two main methods to measure the filament temperature. One is to make the filament also used as a resistance thermometer. This kind of gauge is also called resistance vacuum gauge. The other is to use a thermocouple, also known as thermocouple vacuum gauge.

Normally, thermal conductivity gauge has a measuring range from 10Pa to 0.01Pa, and it can not be used in a state of non-continuous medium state, especially when the pressure is near to 1atm. However, the gas convection and other factors are also taken into account, so that the gauge can measure a vacuum up to 1atm later. Although it is still called thermal conductivity gauge, in reality it is convection vacuum gauge to be discussed below.

(2) Convection vacuum gauge. As the thermal conductivity gauge can not measure the vacuum in a non-continuous medium state, in order to cover the vacuum range from 10Pa to 1atm, Steckelmacher developed a method by

considering natural convection to enlarge the measuring range to 1atm. Because of its focus to extend to 1atm, the accuracy at low pressure is relatively not so satisfactory. At present, the application range of practical convection vacuum gauge is from 0.1Pa to 1atm[3, 4].

In some designs of the thermal conductivity and convection gauges, two thin tubes with the same dimension and structure are both used. One of them is used as measuring tube to measure the vacuum; the other is completely sealed and its internal vacuum is pumped to a certain degree, as a reference tube. This arrangement can eliminate the effects of radiation and axial solid heat losses.

4.2.2 Vacuum gauge problems in use

1) Problems of thermal vacuum gauge

(1) Aging of thermal vacuum gauge. The filament in thermal vacuum gauge is often aging because of the effects of long-term high temperature and gases. So the performances of the gauge will change, including mainly zero drifting and sensitivity changing. There are two types of effects leading to these changes: one happens in the filament itself (such that the filament becomes thin because of oxidation or the lattice change of filament) and the other appears on the surface (such that gas is adsorbed to the surface and an adsorption layer forms so that the surface adaptation coefficient changes which then affect the gas thermal conductivity). The change in filament itself is irreversible; and the emergence of gas adsorption layer is sometimes reversible, sometimes irreversible.

In the pressure range of 0.1 to 10Pa, tungsten or platinum filament is the most easily oxidized. When the pressure is lower than 0.1Pa, gas is rarefied and the filament is difficult to be oxidized; while the pressure is higher than 10Pa, oxidation will be weakened due to good heat transfer that leads to the lower filament temperature. In the same temperature, platinum filament has a much less drift than tungsten.

(2) Thermal vacuum gauge performance curves are different with gas types. Normally, the performance curve measured for air can not be directly used for other gases. Because in a freeze-drying process water vapor accounts for 55%~95% inside the gauge and water vapor has better heat transfer ability than air, the values measured in the cases with high humidity by thermal vacuum gauge are higher than the real values; and generally an average correction coefficient of 0.65 is taken[3].

(3) Thermal vacuum gauges have a higher thermal inertia, so the change of filament temperature along with the pressure variation often has some delay.

(4) Because thermal vacuum gauges are strongly influenced by the external temperature, the vacuum gauge must be installed in some places where the gauge is less susceptible to heat radiant or heat convection.

2) Features of capacitance diaphragm manometer

The diaphragm manometer measures the gas absolute pressure, with fast dynamic time, high measuring accuracy, good repeatability and stability. The measuring result will not be influenced by gas type. The international community has generally adopted this gauge as vice standard gauge and as the transmission standard for international vacuum value intercomparison in 1 to 10^3Pa pressure range. However its price is much higher than thermal vacuum gauge.

4.3 Vacuum pump for freeze-drying

4.3.1 Tasks and characteristics of vacuum pumps in freeze-drying

(1) Removing air and other non-condensable gases leaking into the drying chamber from atmospheric environment.

(2) Removing air and other non-condensable gases released from the materials in the drying chamber.

(3) Keeping the drying chamber and the vapor condenser (cold trap) at low pressure to meet the requirement of sublimation and desorption (generally, the pressure is from 1Pa to 10Pa).

(4) Although in the drying process the water vapor from sublimation and desorption is mainly removed by water vapor condenser; however there is still a little vapor into the vacuum pump. So some measures (such as "gas ballast") should be adopted to solve the problems of vacuum pump.

(5) In the early stage of drying process, a great pumping speed is required; but in the mid and late stages, that the pumping speed is substantially reduced.

4.3.2 Main parameters of vacuum pump [68]

(1) Ultimate pressure: it means the lowest pressure at its inlet which can be achieved by a pump, that is, the lowest pressure when the throughput is zero.

(2) Pumping speed: it refers to the volume flow rate of a pump at its inlet, measured in m^3/h or L/s.

(3) Throughput: it refers to the pumping speed multiplied by the gas pressure at the inlet, and is measured in units of $Pa \cdot m^3/h$ or $Pa \cdot L/s$. It is a parameter that can indirectly reflect the mass flow rate of the gas from the pump.

(4) Starting pressure: it refers to the pump starting pressure with a pumping action. For a pump that can not be directly started under the atmosphere pressure, special attention should be paid to its staring pressure. A vacuum pump's pressure range is from limit pressure to starting pressure.

(5) Compression ratio: it means the ratio of the outlet pressure to the inlet pressure for a given gas in a vacuum pump.

4.3.3 Commonly used vacuum pump combinations for freeze-drying device

Different types of vacuum pumps have a different working pressure range. The vacuum pumps (or pump set) used in freeze-drying devices require that the suction pressure is 1—10Pa and the exhaust pressure is 1atm (about 10^5Pa). Those available vacuum pumps are shown in Table 4.2.

Table 4.2 Available vacuum pumps and their pressure ranges for freeze-drying device

type	application range(Pa)	best application range(Pa)
reciprocating pump	10^2—10^5	10^2—10^5
water ring pump	(2×10^2)—(2×10^5)	10^3—10^5
water ejector pump	(3×10^3)—(2×10^5)	10^4—10^5
multistage steam ejector pump	10^{-1}—10^5	1—10^5
oil seal rotary-vane pump	10^{-1}—10^5	1—10^4
sliding-vane pump	(3×10)—(2×10^5)	1—10^4
roots pump	1—10^5	1—10^4
dry piston pump	(5×10^2)—(2×10^5)	10—10^5
claw type pump	(5×10)—(2×10^5)	10—10^5
scroll pump	(5×10^2)—(2×10^5)	10^2—10^5
screw pump	10—10^5	10—10^5

For small or experimental devices of freeze-drying, only one vacuum pump is used, such as rotary vane oil pump, water ring pump or scroll pump etc., which pumps the gas into atmosphere directly. In order to improve the vacuum, a "two-stage pump" (two stage in series) can be used; the main pump is connected to the vacuum space which directly pump the gas of the vessel to attain the required vacuum.

For medium or large scale freeze-drying devices, pump set is usually adopted to reach a high pumping speed. That pump set takes a singer-stage or a two-stage pump as “backing pump” in its high pressure side and Roots pump as main pump also called booster pump in its low pressure side. In that combination, the exhaust pressure of Roots pump is much lower than 1atm, while its pumping speed (S_{Roots}) is much higher than that of backing pump ($S_{backing}$). The following range can be selected in design.

$$S_{backing} = \left(\frac{1}{3} - \frac{1}{8}\right) S_{Roots} \tag{4.1}$$

The Roots pump consists of two double-lobe (“8”shape) impellers fixed to a pair of parallel shafts. These are rotated driven by gears in opposite directions within the pump housing. Due to the rotation of the impellers, the pumped gas can be drawn to the space between the impellers and pump shell and then discharged through the exhaust. Because the volume of that space will not change, the gas in the pump housing is not compressed. When the space is connected to the exhaust side, there is a back flow of gas from the exhaust region to the space so that the gas pressure rises, but the gas is still forced to be discharged then because of the rotation of the impellers. There is a small clearance between the impellers and the inside wall of pump housing, which makes high speed rotation possible.

Roots pump is characterized by a high pumping speed and the speed can maintain in a wide pressure range. Even if there are some steam, dust etc. in the gas, they would not have remarkable effect on the operation of Roots pump. However the compression ratio of Root pump is quite low, other type of vacuum pump is needed as its “backing pump” when Root pump is used as the main pump.

The ultimate vacuum of Roots pump depends on not only the construction and manufacturing accuracy itself but also the ultimate vacuum of its backing pump. In order to improve the ultimate vacuum of a pump set two Roots pump in series can be used. For medium or large scale freeze-drying equipments, a water ring vacuum pump or an oil sealed mechanical pump with “gas ballast” can be applied generally. Water ring pumps can remove a lot of steam, but its ultimate vacuum is limited by saturated vapor pressure, so it can not be very high. The oil sealed mechanical pump with “gas ballast” can make a higher ultimate vacuum but its ability of removing steam is limited. Therefore, such a pump set can be made in which a single-stage or two-stage Roots pump is used as the main pump and a pump combination is used as a backing pump. The backing pump consists of a water ring pump and an oil sealed mechanical pump with "gas ballast" in parallel. In the early stages of drying, the water ring pump is used to remove the vast gas containing steam. Then the oil sealed mechanical pump is used instead of the water ring pump in the next long time drying process to reach high vacuum.

4.3.4 Principles and function of "gas ballast"

The principles and function of gas ballast in mechanical pump is briefly introduced here. For general mechanical pump (such as oil seal rotary vane pump), the pumping process is a compression process. When a rotary pump is set to pump a gas containing condensable vapor, (for example, water vapor), the vapor is compressed and condensed when its pressure is increased. The liquid water mixes with the pump oil, and as the oil circulates in the pump, the contaminating liquid water will evaporate at the low pressure side of the pump. It limits the attainable vacuum, and also makes the oil emulsified and deteriorates the sealing performance of the oil.

For example, when the working temperature of a vacuum pump is 70℃, the corresponding saturated vapor pressure is 31.2kPa from Table 2.2. The partial pressure of water vapor in the exhaust region should be kept lower than that value. If the partial pressure of water vapor in the drying chamber is rather high, the pressure after compression may be higher than its corresponding saturated vapor pressure and then the vapor condensation happens. In order to avoid this phenomenon, a gas ballast valve has been designed to admit a controlled and timed amount of air into the compression stage of the pump. This extra air is arranged to provide a compressed gas-vapor mixture, which reaches the ejection pressure before the condensation takes place.

4.3.5 Requirement on pumping speed in a freeze-dying process

The performance in a freeze-drying process is related with the conditions of the drying chamber and the materials to be dried, mainly in the following four factors:

(1) The volume V, which includes drying chamber, vapor condensation chamber and the connecting passage.

(2) The sealing condition of the drying chamber and vapor condensation chamber, which can be characterized by leaking rate. On a general vacuum, the ratio of the ambient atmosphere pressure to the drying chamber pressure has exceeded the critical pressure ratio. For the present level of sealing

technique, a leak rate lower than 1 Pa·L/s can be guaranteed [3]. If the pressure in the drying chamber is 1Pa, the required pumping speed of the vacuum pump could be only 3.6m^3/h. On the condition of pressure 1atm and temperature 300K, the air density is about 1.183kg/m^3. Then at the pressure of 1Pa, the mass flow rate corresponding to 1 Pa·L/s is about 1×10^{-5}g/s.

(3) The air mass in the void of the materials (pharmaceuticals and food) and the air mass dissolved or adsorbed in the materials. At atmospheric pressure, there are about 20—30 milliliters of dissolved air per liter of water [4]; while the air quantity in the void of materials is related to the voidage.

(4) The gas released from the construction of materials of the drying chamber and the water vapor condenser. In general conditions, this factor could be ignored.

The requirement on pumping speed in different stages is analyzed as the following.

(1) In the initial pressure decreasing stage, the main task of vacuum pump is to reduce the gas pressure in the drying chamber, water vapor condensation chamber and the connecting passage (the total volume is denoted as V). In general conditions, the later three factors just mentioned can be ignored in this initial stage, and the situation can be simplified as a model of pumping gas from a closed vessel (V). Supposing the pumping speed of the vacuum pump is S, the process time is τ and the pressure in the vessel is P, the following differential equation is established.

$$d\ln P/d\tau = S/V \tag{4.2}$$

The initial condition: $P = P_0 = 1\text{atm}$, when $\tau = 0$.

When the pressure is reduced from the initial P_0 to a certain pressure P, there is a relation between pumping speed S and the process time τ

$$S = \frac{V}{\tau}\ln(\frac{P_0}{P}) = 2.3\frac{V}{\tau}\lg(\frac{P_0}{P}) \tag{4.3}$$

For example, the volume of the drying chamber is 0.7m^3 and the volume of vapor condensation chamber and connecting passage is 0.3m^3, the total volume V=1m^3. If we hope that the pressure could be reduced from $P_0 = 1\text{atm} = 1.0133\times10^5\text{Pa}$ to $P = 1\text{Pa}$ in 5 minutes, the needed pumping speed can be estimated as follows.

$$S = 2.3\frac{V}{\tau}\lg\left(\frac{P_0}{P}\right) = 2.3\frac{1\text{m}^3}{5\text{min}}\lg10^5 = 138\text{m}^3/\text{h} \tag{4.4}$$

(2) In the main stage of drying, the main task of vacuum pump is to remove the air in the void of the materials as well as the air dissolved in the materials

or adsorbed to the materials, which is the third factor mentioned before. The air quantity in the void is related to the voidage of the materials. For example, a material of 10 liters is placed in the above-mentioned 0.7 m^3 drying chamber. If the material has air of 80% of the material's volume, the air quantity contained in the materials at 1atm pressure is about

$$M_{air} = 0.01m^3 \times 0.8 \times (1.18kg/m^3) = 0.00944kg = 9.44g$$

If the pressure in drying stage is 1Pa and the process time is 8 hours, the required air exhaust quantity per hour is $9.44g/8h = 1.2g/h$. At 1Pa pressure, the air density is about 1.183kg/m^3, so the said pumping speed approaches to 100m^3/h.

In a general design, the selected pumping speed is much higher than the above estimated value.

4.4 Refrigeration system and water vapor condenser in freeze-drying

4.4.1 Refrigeration system in freeze-drying

A schematic diagram of a freeze-dryer is given in Fig.4.1. The refrigeration system has two tasks: one is to supply refrigeration to a water vapor condenser to form "cold trap"; the other is to supply refrigeration to the freeze-drying chamber to freeze or vitrify the materials.

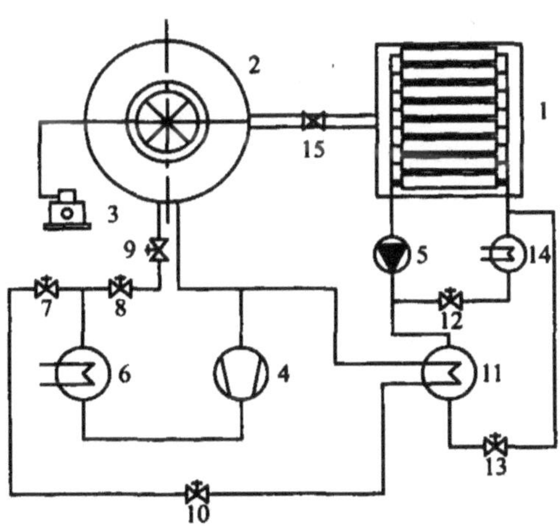

Fig.4.1 Schematic diagram of a freeze dryer system

1. freeze-drying chamber; 2. water vapor condenser; 3. vacuum pump set; 4. refrigerating compressor; 5. circulating pump heat (or cold) carrying agent; 6. refrigerating condenser; 7, 8, 12, 13. electromagnetic valve; 9, 10. expansion valve in refrigeration system; 11. refrigeratiing evaporator; 14. electric heater; 15. vacuum valve

In a freeze-drying system for food products, the temperature of "cold trap" (water vapor condenser) is generally from −40℃ to −30℃, while for pharmaceutical products the temperature is from −80℃ to −40℃.

If the freezing stage is done in freeze-drying chamber, the cooling capacity of the refrigeration system should be able to cool the material to the required temperature. And in the drying process, heating and drying should be guaranteed at low temperatures. Besides, the cooling capacity of the water vapor condenser should be enough to make all the water vapor from the material condense into liquid water as quickly as possible.

There are many ways to realize the refrigeration requirements. For example, liquid nitrogen can be used to reach to −196℃ and it has a large cooling capacity, but there is some trouble in use. By adopting multistage thermo-electric refrigeration (Peltier effect), the temperature can reach to −40℃ or lower, but the cooling capacity is relatively low, so it is just used in experimental freeze-dryer at present.

Vapor compression refrigeration system is widely used in freeze-drying equipments at present, which is shown in Fig.4.1. It mainly consists of a refrigerating compressor 4, a refrigerating condenser 6, expansion valves 9 and 10 and evaporators 2 and 11. There are two refrigeration evaporators and two expansion valve in the system. They provide cold supply separately for "cold trap" 2 and freeze-drying chamber 1. The electromagnetic valves 7 and 8 are used to switch the refrigerant circuit. The evaporator is a surface type heat exchanger with an evaporator inside to evaporate the refrigerant and a condenser outside to condense the water vapor in the vacuum system. The heating and cooling of the freeze-drying chamber are completed by a heat carrying agent (silicon oil), which is heated by electric heater, of cooled by refrigerant.

In order to achieve the required temperature, the vapor compression system can adopt a single refrigerant and two-stage system or a binary refrigerant cascade refrigeration system or a multi-component refrigerant auto-cascade refrigeration system.

4.4.2 Water vapor condenser in freeze-drying

The water vapor condenser, also called "vapor captor", has a task to trap the water vapor sublimed and desorbed from the materials and make the water into ice by desublimation. Then the ice will melt and be drained off. In this process, water vapor should be prevented from entering the vacuum system as much as possible.

In a freeze-drying system, the volume flow rate of water vapor is very large. It is known from Table 2.2 that at 1Pa pressure (the corresponding saturated temperature is about -60℃) the specific volume of water vapor is up to 9.1×10^4 m^3/kg. If there are 10 kilograms of water contained in the materials and the required drying times are 8 hours, the average pumping speed will be up to $1.14\times10^5 m^3$/kg. A general mechanical vacuum pump is hard to achieve such high pumping speed and the vacuum pump can not afford so much water vapor. Sometimes, vapor ejector can be up to this required pumping speed of water vapor, but its ultimate pressure can not satisfy the requirement of freeze-drying.

Therefore, the large amount of water vapor can be only removed by water vapor condenser. The condenser should not only reach to a very low temperature, but also have enough cooling capacity and heat transfer area. According to Table 2.2, at 1 Pa pressure (the corresponding saturated temperature is about -60℃) the sublimation latent heat is 2836.27 kJ/kg. If the desublimation latent heat is also estimated according to that value, then in order to trap the water vapor at -60℃ with a mass flow rate of 1kg/h, the refrigeration system is required to be able to provide about 0.79kW cooling capacity at a temperature lower than -60℃. This cooling capacity is quite high, because for the refrigeration system the cooling capacity will rapidly decrease with the decrease of evaporation temperature.

The water vapor sublimated and desorbed from materials is converted into ice by desublimation on the condenser; it is an unsteady process. In the initial stage of sublimation, there is lots of water vapor sublimated; at that time, thick ice layer has not formed on the outside surface of condenser yet, therefore the desublimation of the water vapor is very quick due to the small thermal resistance. After a period of drying, the drying layer of materials becomes thicker, the vapor escaping resistance increases, and the vapor mass flow rate becomes lower. Meanwhile, the ice layer on the surface of condenser becomes thicker and the thermal resistance increases. So, the desublimation rate of the vapor on the condenser surface decreases.

At 1Pa pressure (the corresponding saturated temperature is about -60℃), the density of ice is

$\rho = 0.925\times10^3 \text{kg}/\text{m}^3$, the sublimation latent heat is $r = 2836.27 \text{kJ}/\text{kg}$

and the thermal conductivity is $k = 2.91\text{W}/(\text{m}\cdot\text{K}) = 10.48\text{kJ}/(\text{m}\cdot\text{K}\cdot\text{h})$ [6, 13].

(In some literatures, the thermal conductivity of ice is regarded as that of frozen food, such as the value $k = 6.3\text{kJ}/(\text{m}\cdot\text{K}\cdot\text{h})$. In fact the thermal conductivity of frozen food is much lower than that of ice[6])

For 1m^2 outside surface of the condenser, if the thickness of ice layer reaches to $\delta = 1\text{cm} = 0.01\text{m}$, the desublimation capacity is about

$$G = F \times \delta \times \rho = 1\text{m}^2 \times 0.01\text{m} \times 0.925 \times 10^3\,\text{kg}/\text{m}^3 = 9.25\text{kg} \qquad (4.5)$$

If the evaporation temperature of the refrigerant inside the condenser remains unchanged, the thickness of ice will become thick and the thermal resistance will increase. As a result, the desublimation temperature will increase and the speed of sublimation will slow down. Generally, an ice layer thickness exceeding 1cm is not expected. For a too thick ice layer, "ice melting" (commonly called as "defrost") should be performed in time. Defrost in the drying stage may affect the quality of freeze dried products, therefore, the area of condenser should be large enough and its construction design needs to be considered carefully, so that the ice layer thickness on the most outside surface of the condenser is basically uniform.

The desublimation rate is related to several factors, which include the vapor pressure on the surface of condenser, the content of water vapor and the thickness of ice layer, as well as the temperature refrigerant (or secondary refrigerant) inside the condenser. The temperature drop across the ice layer is a part of the total temperature difference between the outside gas and the inside refrigerant of the condenser. The lower the refrigerant temperature, the larger the total temperature difference and the higher the desublimation rate.

The thermal resistance of ice layer is one part of the total heat transfer thermal resistance. If the thickness and the temperature difference between the two sides of ice layer are given, the heat flow rate of condenser can be estimated by the heat conduction of the ice layer.

At 1Pa pressure (the corresponding saturated temperature is about −60℃), if the thickness δ of the ice layer is 1 cm and the temperature difference Δ*T* of the two sides is 2.7 K, the heat flow rate is

$$q = \kappa\frac{\Delta T}{\delta} = 10.48\frac{\text{kJ}}{\text{m}\cdot\text{K}\cdot\text{h}} \times \frac{2.7K}{0.01\text{m}} = 2830\text{kJ}/(\text{m}^2\cdot\text{h}) \qquad (4.6)$$

In this condition, the desublimation is

$$\frac{q}{r} \approx 1.0\text{kg}/(\text{m}^2\cdot\text{h}) \qquad (4.7)$$

If the situation is changed as: If the thickness δ is 0.8 cm and the temperature difference ΔT is 5 K, the heat flow rate becomes 6550 kJ/(m^2·h) and the desublimation rate becomes 2.3kg/(m^2·h).

The above discussion is about the desublimation of water vapor. If during the desublimation the water is not converted into ice but frost, the heat transfer condition will worsen considerably. This is because the thermal conductivity of frost is much lower than that of ice (only one tenth of ice) and the density of frost is much less than ice so that the frost layer will be very thick.

4.5 Small-scale freeze-drying equipment for laboratory use

Intact freeze-drying equipment is a set of complexity installation, which has functions such as vacuum pumping, refrigeration, heating drying, control, cleansing and disinfection in place. The freeze-dryers can be classified in accordance with different ways. According to the freeze-drying objects, they can be divided into pharmaceuticals and food freeze-dryers. According to the equipment running ways, there are intermittent freeze-dryers and continuous freeze-dryers. According to the processing capacity, there are large, medium and small scale freeze-dryers. According to the application occasion, there are experimental application, trial application (products intermediate trial application) and production application freeze-dryers.

The area of the internal shelf inside freeze-dryers, for experimental type, is generally about 0.1 m^2. For the trial application freeze-dryers, the shelf area is below 1 m^2. For the pharmaceuticals production freeze-dryers, the shelf area can be different from a few to several tens of square meters. For the food production freeze dryers, shelf area can be as large as several hundreds of square meters.

Experimental freeze-drying devices are mainly used for the exploration of freeze-drying characteristics of new materials, and for the identification of freeze-drying process technological parameters of new materials, in order to achieving high quality freeze-dried products, and reducing the time and energy consumption.

Small-scale laboratory freeze-drying equipment has some finished products, and also has many devices designed by researchers. Such as the FreenZone freeze dryers of Labconco Company in United States, VirTis freeze dryers produced by VirTis U.S. Company, LGJ-10 type machine produced by Beijing Sihuan scientific instrument factory, LG-5-type freeze dryer of Shanghai Centrifuges Research Institute, etc. There are also some small freeze-drying equipment systems developed by us and other researchers [4].

Here a brief introduction of FreenZone 2.5 liter type desktop freeze dryers produced by Labconco Company of the United States is given. The system consists of four parts: vacuum pumps, drying room, refrigerator and the cold trap. A photo of the freeze dryer is shown in Fig. 4.2. Refrigeration system uses two-stage cascade vapor compression refrigeration cycle. It uses two hermetic compressor, whose power are respectively for 1 / 3 and 1 / 4 hp. Air-cooled heat exchanger is adopted in this refrigeration system. Cold trap is made of stainless steel, and its cold trap temperature can reach −84℃; Catching water capacity of the cold trap is 1.25 liters / 24 h; ice condensate capacity is 2.5 liters. The vacuum pump is the British Edwards rotary vane vacuum pump, its pumping speed is 117 L/min, and the ultimate pressure is 0.1Pa. A dry filter and organic solvent filter are equipped in front of vacuum pump. Vacuum connection uses international standard pieces, in order to make the dismantling and installation simple and convenient, sealing performance reliable. The drying chamber of the freeze dryer is consisted with the upper cover plate, main plate, shelf, cylinder and lower cover plate. Drying chamber is the transparent bell-jar type. Stainless steel sample frame is specially designed, and the sample plate spacing is adjustable. There is manual stopper organization, which can make good sealing of bottles or vials in the drying chamber. There are two heating shelves, which use electric heating to reach a stable temperature. There is a specially designed drain valve at the bottom of the drying chamber, which is safe and convenient. Operating control panel has LCD digital display. It can display cold trap temperature, vacuum degree, alarm, and can also provide computer communications interface. (http://www.labconco.com/-scripts)

FreenZone bench top freeze dryers have a shortage that they do not have cooling device for freezing material in drying chamber. Therefore, the freezing or vitrification of the materials must be completed outside the drying chamber, and then the materials are moved into the shelf of drying chamber. In the drying chamber, external radiation heat is very large, the material temperature will rise obviously during pressure decreasing process. If the material temperature is higher than its glass transition Temperature or eutectic temperature, it will lead to the material collapse or melting. This will be a huge impact to the quality of the freeze-dried products. To solve these problems, some mini changes to a few parts of the freeze dryer have been tried, which can achieve the freezing of materials in the drying chamber [4].

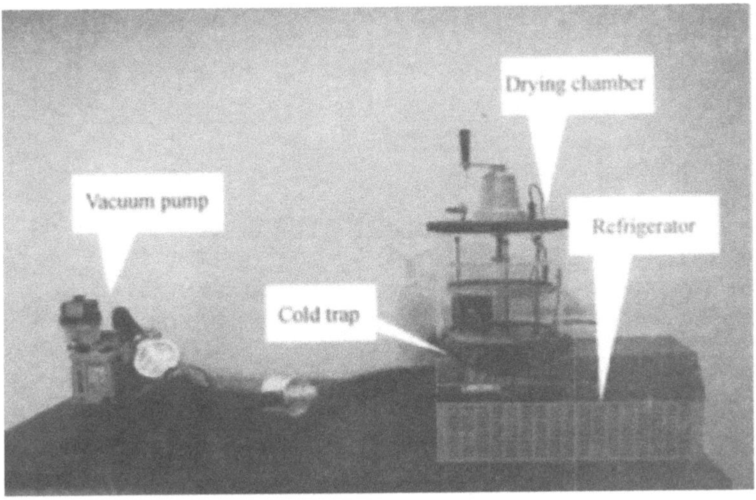

Fig.4.2 FreeZone 2.5 Liter Experimental Freeze Dryer of Labconco Corporation

4.6 Intermediate experimental freeze-drying equipment

Intermediate experimental freeze dryers are mainly used in a variety of pharmaceuticals and food research institutions; intermediate experiment is a technological exploration before mass production. In general, the technological data obtained from exploration on intermediate experimental freeze dryers, after making minor adjustments to process parameters according to running situation, can be directly used in the production freeze dryers. Therefore, the requirement for direct enlargement is rather high, that is, all the structure characteristics and process parameters in the intermediate experimental freeze

Fig. 4.3 An intermediate experimental freeze dryer

dryer should be the same with the production freeze dryer, except the different sizes.

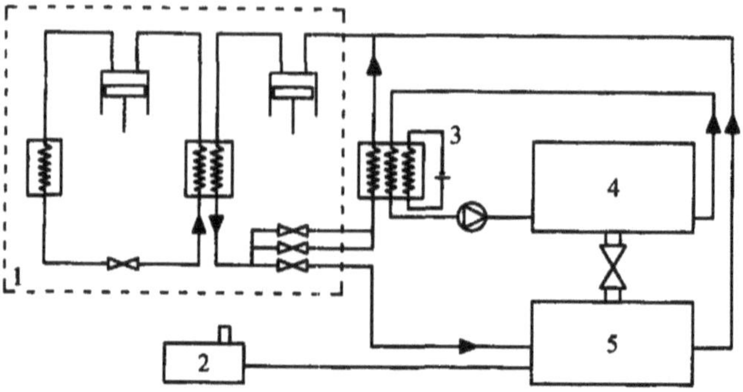

Fig.4.4 System diagram of an intermediate experimental freeze dryer
1. refrigeration system; 2. vacuum system; 3. heating and cooling heat exchanger; 4. drying chamber; 5. cold trap

Intermediate experimental freeze dryers supplied currently by the market, mainly use single-stage vapor compression refrigeration system, and the refrigeration temperature can minimum up to about −45℃. This temperature can not meet the requirement of glass transition temperature for many pharmaceutical materials. In order to achieve a lower temperature, Shanghai University of Science and Technology and Shanghai Tofflon Company have cooperated in the development of the intermediate experimental freeze dryers using vapor compression cascade refrigeration system, so that the lowest refrigeration temperature up to about −80℃. The freeze-drier device is shown in Fig.4.3. This device is mainly constituted by the drying chamber, cold trap, refrigeration system, vacuum system, and heating system as shown in Fig. 4.4.

This refrigeration system is a cascade refrigeration cycle which is composed of two parts: high temperature stage and low-temperature stage. The two-stage system uses an evaporative condenser to link up. The evaporative condenser is not only the evaporator of high temperature stage, but also the condenser of low temperature stage. The condenser of high temperature stage can be water-cooled or air-cooled. The power and heat dissipation capacity of intermediate experimental freeze dryer are not very large. For easy usage, air-cooled condenser is usually adopted.

Cooling and heating of shelves use silicon oil as cooling (heating) carrying agent. In Fig.4.4, Part 3 is a heat exchanger to cool (or heat) the carrying agent. The agent is cooled by the refrigeration system, or heated by electricity. The agent (silicone oil) is driven by a circulating pump to the shelves, for cooling (or heating) them.

High pressure liquid refrigerant in low temperature stage is divided into 3 lines. One line, through the expansion valve, directly provides cooling to the low temperature cold trap (5) by direct evaporation. The second line, through the expansion valve, enters into a heat exchanger (3) for carrying agent. The third line, through other expansion valve called "maintaining cold valve", enters into the agent heat exchanger (3). The role of "maintaining cold valve" is to maintain the shelf temperature at a certain low temperature range in the drying process. When the shelf temperature is higher than the setting value, the "doped cold valve" is opened, and provide more cooling to the shelf, through the valve.

Vacuum system uses Japan's ULVAC two-stage rotary vane vacuum pump, which can make the ultimate pressure vacuum less than 1Pa in the drying chamber. The technical characteristics of this intermediate experimental freeze dryer are shown in Table 4.3.

Table 4.3 Technical characteristics of an intermediate experimental freeze dryer

effective area of shelf	$0.18m^2$
shelf size	300mm×300mm×16mm
shelf number	2+1
spacing between shelves	90mm
Total power	4kW
Total weight	800kg
Overall dimensions	1420mm×1000mm×1800mm
capacity of water capture	4 kg
loading for 22mm vials	400 vials
loading for 16mm vials	700 vials

4.7 Intermittent and continuous freeze-drying equipment

Freeze-drying equipment, according to equipment operating mode, can be divided into intermittent freeze-drying machine and continuous freeze-drying machine. Food Freeze Dryer both has intermittent and continuous types, but pharmaceutical freeze-drying machine is mainly intermittent.

4.7.1 Intermittent freeze dryer

An intermittent freeze dryer, called also bath type freeze dryer, is shown in Figure 4.1. The advantages of intermittent freeze-dryers are:

(1) It is suitable for a multiple variety, middle or small batch pharmaceuticals and food production, especially for seasonal food production.

(2) It is single machine operation, if a piece of equipment failures, it will not affect the normal operation of other equipment.

(3) It is convenient for the equipment manufacturing and maintenance.

(4) It is convenient to control the heating temperature and vacuum degree at different stages.

Its shortcomings are as follows:

(1) As a result of preparatory operation, such as loading, unloading and start-up process, the equipment utilization is very low, and the energy consumption is very high.

(2) For large production requirements, multiple single-machines are required, and each single-machine must be accompanied by a subsidiary system. It causes a quite high equipment investment and operating cost. At present, advanced batch type freeze-dryers have advanced centralized control system, which can realize sequence startup or alternation work to multiple drying chambers, and optimize the multi-unit system, thus, it can improve equipment utilization and save energy consumption.

4.7.2 Continuous freeze dryer

A diagram of continuous freeze dryer is shown in Fig. 4.5. It is suitable for a single species, large production and adequate raw materials products, particularly for paste or particle food production. The advantages of continuous freeze-dryers are higher equipment utilization, convenient for automate production. The drawbacks of continuous freeze-dryers are that the equipment is huge, complex and difficult for manufacture. The vacuum sealing problems of the loading and unloading ports require especially a higher level technique.

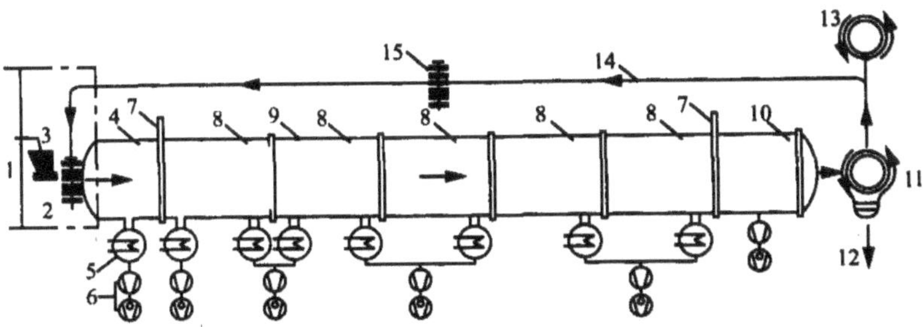

Fig.4.5 Diagram of a continuous freeze dryer

1. freezing chamber; 2. loading room; 3. tray filler; 4. loading isolation room; 5. water vapor condenser; 6. vacuum pumping system; 7.gate valve; 8. freeze drying tunnel; 9. heating plate with the lifting and transport device; 10. discharge isolation room; 11. discharge room; 12. product export; 13. cleaning device; 14. crane rail for sending transport; 15. lifting transporter

4.8 Food freeze-drying equipment

Compared with the freeze dryers of pharmaceuticals, the freeze dryers of food production have following distinct features. It has a large processing capacity, and it does not need very strict process parameters; Cold trap temperature and shelf temperature do not need very low; and the heating temperature of secondary drying could be higher.

Using TFD-2000LF2 freeze-dryer as an example, which is manufactured by Japan Toxyox Technical Research Kabushiki Kaisya, the characteristics of food production freeze-dryers are discussed. It has characteristics such as large dealing amount, short drying time, low energy consumption, reliable operation, high automation level, simple operation, easy to maintenance, etc., so it is suitable for mass production of various types of food [69]. A brief introduction of the main components is followed.

4.8.1 Vacuum system

The vacuum system uses a combination of two pump sets, as shown in Fig.4.6, rotary vane pump set (A, SR-75BI, SHINKO SEIKI) and combined vacuum pump set (B, FT4-450, ANLET Corporation). When the material is loaded, the pump set (A) is running to meet the large pumping speed in the early stage of drying. Then the combined vacuum pumps (B) system is operated, to maintain the system at 40—10 Pa and it can reach below 10Pa in latter stage. This combined working mode can rapidly improve vacuum degree, and can reduce energy and time consumption.

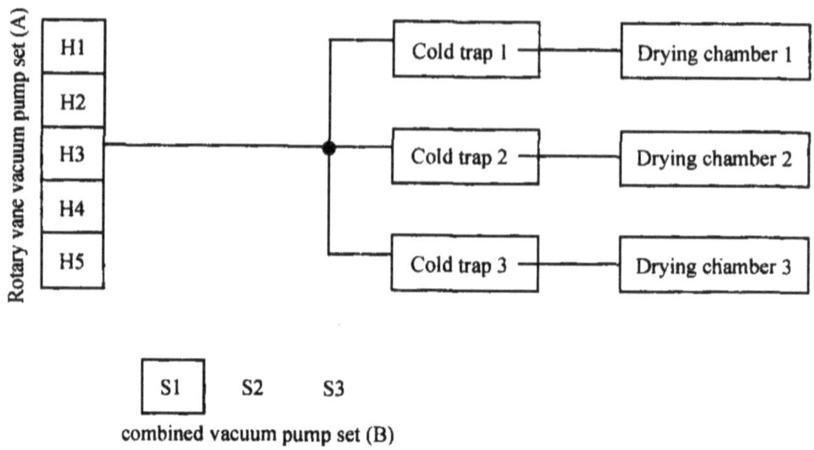

Fig.4.6 TFD-2000LF2 type freeze dryer system for food production

4.8.2 Heating system

1) Drying chamber

The loading materials trays are put on the loading trolley (Fig.4.7). They enter firstly a quick freezing room through rail, and then loaded to drying chambers. The drying chambers are horizontal cylinder type, with diameter of 2.3 m, and length of 11.8 m. There are two rows of shelves in the drying chamber, each raw has 17 layers (Fig.4.8). The trays with frozen food are loaded into the spaces among the shelves using loading platform car through rail. The food in the trays is heated by the upper and lower shelves. Cold trap is at the side of the drying chamber, 9 m in length.

Fig.4.7 Loading material frame of food freeze-dryer

Fig.4.8 Drying chamber of food freeze dryer

2) Heating plate and heating carrying agent heat exchanger

In some food freeze-dryer, distilled water is used as heating carrying agent. The distilled water is heated by high-temperature steam; then the distilled water is used to heat shelves. TFD-2000LF2 freeze dryer, uses silicon oil as heat carrying agent. High pressure steam produced by boiler, through the heat exchanger (plate heat exchanger, model M10-MFG), is used to heat the silicon oil, as shown in Fig. 4.9.

Silicone oil is very stable, even for long-term usage it does not cause corrosion to the shelves. The flow section of the heating carrying pipes is very large, the resistance is small, and the silicon oil is full of the pipes. Its heating temperature can reach 100 ℃, and the temperature difference between heating carrying agent and shelf's surface is quite small.

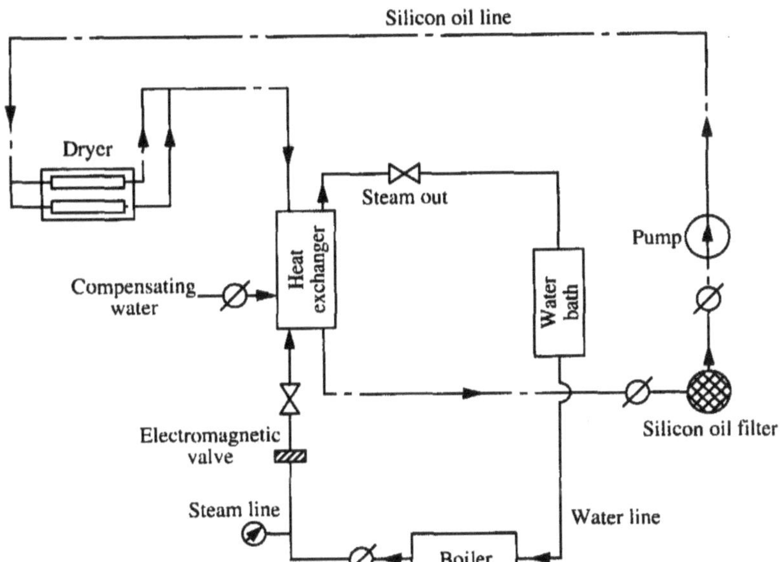

Fig.4.9 Heating system of a food freeze-dryer

In the early stage of drying process, in order to speed up the drying, the materials on the shelves are rapidly warmed to achieve the allowed maximum temperature, and then adjust the temperature to prevent materials overheating. According to the temperature history curve required of the program regulator, the temperature program control is carried out.

When shipped out from the freezing room, the frozen food temperature is about −25℃. After entering into the drying chamber, the pressure is firstly vacuumed to about 80—90 Pa, again to about 40—10 Pa. At that time, the temperature is about −30℃. Sublimation drying is carried out for about 11 h at −30℃ (at this stage, the silicone oil temperature is in the range of 100℃ and 80℃ for regulation). When the material temperature curve noticeably rises, the silicone oil temperature must be regulated, to make the food temperature rise and then maintain at 60—70℃, for the desorption drying, it lasts about 6—7h.

4.8.3 Refrigeration system

1) Vapor compression refrigeration unit

The refrigeration system is composed of 2 sets of 55kW and 1 set of 37kW of dual screw double stage compressors. The output power of the compressors is automatically regulated to meet the load variation of cold trap.

2) Technical parameters of refrigeration system (Table 4.4)

Table 4.4 Technical parameters of refrigeration system in TFD-2000LF2 food freeze dryer

item			units	types	
model				RUS-2004TF	
refrigeration capacity			kW	269.2	
operating conditions	condensing temperature		℃	40	
	evaporating temperature		℃	-30	
	cooling water	inlet	℃	32(max)	
		outlet	℃	37(max)	
compressor	compressor Model			7500SU-T	5000SU-T
	number of compressor			2	1
	rotational speed		r/min	2800	
	low stage output		m^3/h	589.4	208.7
	high stage output		m^3/h	230.8	95.0
capacity control method			—	Sub-control	
capacity control range			%	100, 75, 56, 38, 19, 9	
motor	output power		kW	55×2+37	
	input power		kW	136.2	
	operating current		A	243	
	power Supply		—	AC3 Φ, 380V, 50Hz	
	startup way		—	sequential startup	
condenser of the refrigeration system	the maximum pressure		MPa	0.7(in cooling water side)	
	flow rate of cooling water		m^3/h	61.2+(8.5)	
	head loss		kPa	below 100	
	refrigerant volume inside the condenser		liter	648	
refrigerant			—	R22	

3) Cold trap

Water vapor cold trap set (FD214) is configured by the combination of three cold traps. The cold trap set can be used alone, and also can be used alternately or simultaneously. This makes it easy for a separate defrosting, easy for separate maintenance. It does not need downtime maintenance, and is suitable for continuous usage. The catching water capacity of the cold trap set is 2000 liters each batch.

The cold trap has non ferrous metal surface which is low temperature tolerance, with the characteristics of heat transfer quickly, strong catching water capacity, corrosion resistance, long service life and so on. The cold trap temperature of food freeze dryer is generally −30℃ to −40℃.

4) Melting ice (Defrosting) system

Defrosting system is constituted by defrosting cans with built-in vapor heater, drainage pump and various valves, a level gauge, temperature sensors and so on. When a cold trap is turned into defrosting mode, a defrosting can is connected with the cold trap to make the frost on the cold trap melt into water, and discharged from the vacuum pump.

4.8.4 Control system

Using industrial computer automatic control system, this system has the functions of programming control, touch screen display and setting control command. Fig.4.10 is the photo of the touch-control panel.

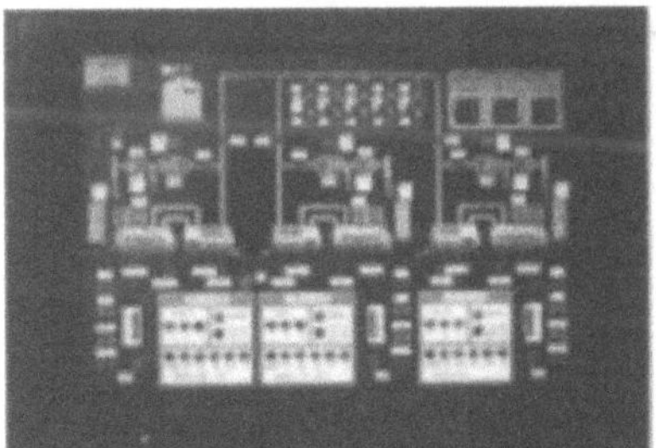

Fig.4.10 Touch control panel of TFD-2000LF2 freeze dryer

4.9 Pharmaceutical freeze-drying equipment

4.9.1 Characteristics of pharmaceutical production freeze dryer

Compared with intermediate experimental freeze dryer, freeze drying system for pharmaceutical production, not only has the drying chamber, vapor cold trap (condenser), refrigeration system, vacuum system, control system, but also has the hydraulic system, CIP systems (Cleaning in place, online cleaning system), SIP system (Sterilization in place, online Sterilization System) and safety interlocking device etc.

A freeze dryer of pharmaceutical mass production, Lyo-7.5 type freeze dryer as an example, is shown in Fig.4.11, and the shelves and trays of the freeze-dryer are shown in Fig.4.12[70].

Fig.4.11 Lyo-7.5 pharmaceutical production freeze dryer appearance

Fig.4.12 The shelves, trays and pharmaceuticals inside Lyo-7.5 production freeze dryer

Drying chamber of the freeze dryer is a rectangular box, made of AISI316L stainless steel, and surface roughness *Ra* <0.75μm. There are indirect heating and cooling shelves (seven) in the drying chamber, which are made of AISI316L stainless steel. Shelf group is installed in the freeze-drying box by frame and slide, and is driven by hydraulic pistons for the moving up or down. The thickness of shelves is 20mm, and there are flow channels inside the shelf for silicon oil. The temperature distribution inside the shelf is uniform, within

± 1℃. The flatness of shelf is ± 1mm / m or less; its surface roughness *Ra* <0.75μm, these are all in line with the GMP requirements.

On the top of the drying chamber, there are hydraulic cylinder, thermal vacuum gauge probe, connection flange, discharge valves, limited leakage valve and verification holes.

Cold trap (condenser) has coil structure, whose material is AISI321 seamless stainless steel pipe. Spray / immersion method is used for defrosting. Temperature probes are 2 Pt-100 platinum resistances, which are inserted into the measuring hole of the cold trap coil through the blind pipe.

The main part of refrigeration system is two stand-alone two-stage compressors. Portion of the direct expansion refrigerant goes into the cold trap, and the other portion goes into the heat exchanger to heat silicon oil.

4.9.2 System of pharmaceutical production freeze-dryer

The total system of Lyo-7.5 freeze-dryer is shown in Fig4.13; it consists of eight subsystems as following.

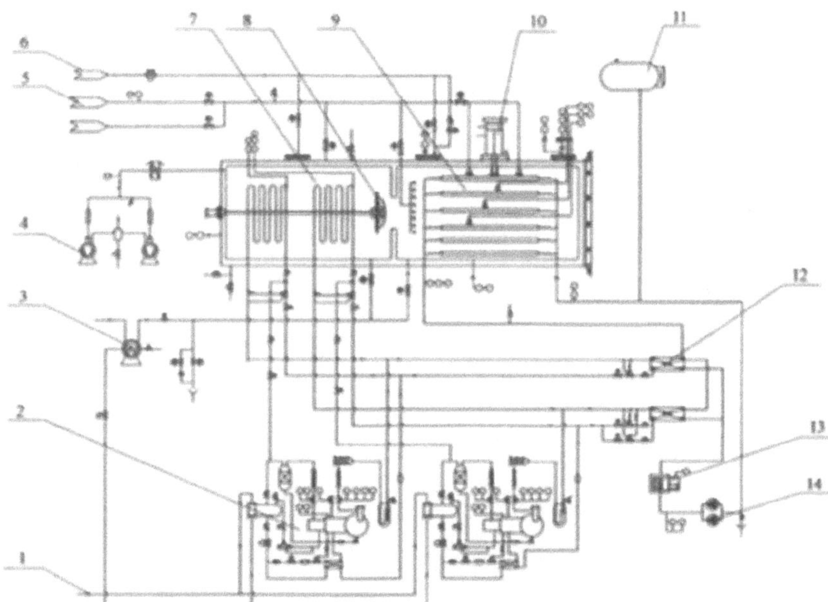

Fig.4.13 Drugs production freeze dryers system

1. cooling water; 2. condensing unit of refrigeration (2 sets); 3. water ring vacuum pump; 4. rotary vane vacuum pump; 5. cleansing water; 6. dry air; 7. cold trap; 8. diaphragm valve; 9. shelf in drying room; 10. hydraulic system; 11. expansion box; 12. heat exchanger; 13. heater; 14. circulating pump

1. Refrigeration subsystem

Refrigeration subsystem consists of two sets of two-stage compressor unit. Each refrigeration units can be independent cycle, that is, failure of any one set of units, the system can work as usual. The cold trap cooled by refrigerant direct expansion, the limit temperature is −70℃. The shelves in the drying chamber are cooled by secondary refrigerant (silicon oil); the shelf temperature can reach −55℃; In the sublimation process, the system will adjust the materials temperature according to technical requirement.

2. Vacuum subsystem

The vacuum system is composed of a water ring vacuum pump set and a rotary vane vacuum pump set. The drying chamber must maintain a certain vacuum degree, and the vacuum system must provide adequate pumping speed. Vacuum degree of the vacuum system should be matching with products sublimation temperature, and cold trap temperature; too high or too low vacuum degree is not conducive to drying. The vacuum pump set in the freeze-dryer has sufficient pumping speed to ensure that it can reach the required vacuum (its ultimate pressure is less than 1 Pa).

3. Control subsystem

Control system is mainly composed of the touch screen and Programmable Logic Controller (PLC), which can be operated automatically or semi-automatically, and can also be manually control. The valves used in the system are pneumatic valves. Control contents of this control system include freezing, vacuum, leakage, heating, cold-doped, exhausting, and defrosting (frost), etc., and have comprehensive alarming and linkage protection devices, in order to avoid misoperation. The control screen shows the real-time operating status of major components. Processing data can be displayed on the touch-screen and the logger; and it can also be stored, to prepare for calls.

4. The carrying agent circulatory subsystem

The cooling/heating carrying agent is circulated by a shielding stainless steel liner catamaran circulating pump.

5. Hydraulic subsystem

Hydraulic system is a combination of the pumping station, the oil cylinder and hydraulic parts. It generates two independent hydraulic circuit, one circuit is used for the oil tank at the top of freeze-dried room, to move the shelves and for convenient to cleaning work, and to drive the stopper at the end of drying, as shown in Fig.4.14. The other circuit is used to control the opening and closing of the diaphragm valve (mushroom valve, No.8 in Fig.4.13).

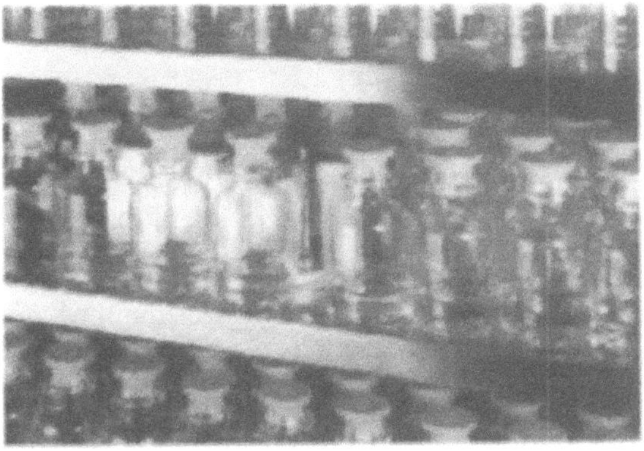

Fig.4.14 Hydraulic devices for stopper

6. CIP subsystem (cleaning in place system or online cleaning system)

The online cleaning and online sterilization of freeze-dryer in discussed in Chapter 8 in this book. Online cleansing is also known as CIP (cleaning in place). The usual procedure of CIP is to use reverse osmosis water cleaning 5min, reuse injection water cleaning 5min. After online cleansing, its cleaning effects should be corresponded with the GMP requirements.

In order to achieve online cleansing and online sterilization, the freeze-drying chamber and the cold trap are equipped with injection rods and high-pressure nozzle. Rotating airflow style nozzle is shown in Fig.4.15. If the pressure is higher than 0.3 MPa, cleaning water mist is sprayed through the nozzle; at the same time, the shelves moving up and down, therefore all the corners can be cleaned. After online cleaning, valves are closed, the water ring vacuum pump start to work for drainage, and make a thorough disinfection to the drainage.

Fig.4.15 Rotating airflow style nozzle for online cleaning and sterilization

7. SIP subsystem (sterilization in place)

Freeze-dried drugs are mainly sterile products. In the process of preparation freeze-drying, the freeze-drying chamber and the vacuum condensation systems are all required to be controlled as a sterile space. At present, the high-temperature pure steam is mainly used to sterile, its technical conditions are 121℃ and 30min. It must be carefully considered in the design that the tolerance ability of the parts to high temperature steam, such as the sealing parts in the oil tank and in the vacuum valves etc. Freeze-dried chamber and condenser must be designed and manufactured according pressure vessels standards. Cold trap and freeze-dried chamber are equipped with safety valves, temperature control and maintenance and calibration devices. Steam quality should be corresponded with the injection water (WFI, water for injection) standards. Throughout disinfection process controlled by the PLC and monitored by the software is completely automated.

4.9.3 Main technical parameters of Lyo-7.5 freeze dryer

The main technical parameters of the freeze dryer are summarized in Table 4.5.

Table 4.5 Main technical parameters of Lyo-7.5 freeze dryer

effective shelf area	7.56 m^2
loading for Φ22mm vials	15 000 bottle
loading for Φ16mm vials	31 500 bottle
loading for 2ml ampoules	58 000 bottle
maximum catching water capacity of the condenser	120 kg
equipment specifications	
shelf size (L × W)	1200mm×900mm
number of shelves	7+1 block
spacing between shelves	100mm
shelf temperature (unloaded)	-50 ~ +70℃
condenser temperature (unloaded)	-70℃
system leakage rate	≤1×10^{-2} mbar·(L/s)
capacity of vacuum pump	from atmospheric pressure to 0.1mPa: <30min
ultimate vacuum pressure	<2.6 Pa
requirement	
power	40kW
cooling water(<25℃)	12t/h
compressed air(>0.4 MPa)	0.1 m^3/min
overall dimensions and weight	
$L\times W\times H$	4400mm×1500mm×3000mm
weight	5500kg

5

Freezing-drying of Food

5.1 The characteristics of freeze-dried food

Food freeze drying is a dehydration process in which the water in food is frozen, and then sublimated directly from the solid phase to vapor by heating under vacuum.

5.1.1 The advantages of freeze-dried food

(1) The process at low temperature and low pressure makes freeze drying an effective way to keep the color, smell, flavor and heat-sensitive nutrients of food and also eliminates the surface hardening of food.

(2) Freeze-dried food is porous and easy to be rehydrated and instantly dissolved. It can be consumed directly or after rehydration.

(3) Since freeze-dried food contains very low moisture content, it has relatively small density and is easy to be transported. The freeze-dried food can be preserved at room temperature for a long time, while the cost of transportation is much lower than that of frozen food.

(4) No additives are added into the food during the freeze drying process.

5.1.2 The disadvantages of freeze-dried food

(1) If exposed directly to air, freeze-dried food will be rehydrated quickly and a series of chemical reactions will happen, resulting in the deterioration of food. The freeze-dried products have to be vacuum-packaging or vacuum-nitrogen charged packaging. The packaging materials should be waterproof.

(2) During transportation and sale process, freeze-dried food is easy to be powdered or cracked for its loose porous structure.

(3) Freeze-drying is a time-consuming and energy-consuming process, which lead to higher product costs of freeze-dried food.

(4) Freeze drying system includes vacuum and refrigeration equipments. The initial costs are relatively high.

(5) Freeze drying of food demands higher technological requirements and the techniques are relevant to the types of foods.

Statistically, there is about 20% of food products become deteriorated in China as a result of improper storage and distribution every year. Freeze-drying of food is regarded as a new high-technology of food dehydration. It is being widely used for long-term storage and long distance transportation of foods.

Freeze-dried foods are high value-added products. In international market, the price of freeze-dried food is quadruple of the dehydrated foods. At present, freeze-drying is mainly used to high-quality, instant, leisure foods, vegetables and fruits for export, and foods for tourism, exploration and navigation. The production costs and prices of freeze-dried food will be decreasing along with the improvement of techniques and equipments.

5.2 Techniques for food freeze-drying

The food freeze-drying includes four main processes: preparation and pretreatment, freeze-drying, packaging and storage, rehydration. Of the four, freeze-drying is the key and difficult process

5.2.1 preparation and pretreatment

It is necessary to treat the food by physical and/or chemical methods before freeze-drying process. The pretreatments include sorting, cleaning, slicing, blanching, sterilization, concentrating and so on. The pretreatment methods are different for different types of foods.

1) Pretreatment of fruits and vegetables

Pretreatments of fruits and vegetables include sorting the raw materials, cleaning, peeling, trimming, blanching and so on. Sorting and cleaning are essential processes for all types of fruits and vegetables. The other pretreatment processes are selected depending on the types of raw materials and products.

The purpose of pretreatments is to select high-quality raw materials, to reduce the loss of color, smell, flavor and nutrients of foods during the processing, storage and transportation processes, and to improve the heat and mass transfer efficiency in freeze-drying process.

The sorting of raw materials includes selection and classification. Selection is necessary for raw fruits and vegetables to remove the unqualified, harmful, and deteriorated raw materials before freeze drying. Then the raw materials are classified according to their size, color grading and maturity. The sorting is used to simplify the peeling and blanching process, to reduce the waste of raw materials; and to guarantee the quality of the products.

The fruits and vegetables must be washed before freeze drying to remove the dust, some micro-organisms and pesticide residue. Rinsing is usually applied to clean the fruits and vegetables. Rinsing is performed by running water in the sink, by spraying with water gun; or by the cylinder washer.

The selection of peeling methods depends on the types of fruits and vegetables. The fruits and vegetables that have hard, non-edible and strange smell skin must be peeled. The flesh of fruits should not be injured in peeling process. Peeling should be fast, high efficiency and economy. The peeling methods include mechanical, chemical, thermal peeling and hand peeling. Some peeled fruits and vegetables must be kept from exposing to air to avoid browning.

The conditions of vacuum and low temperature can inhibit the oxydase, peroxidase, nonenzymatic browning of the fruits and vegetables. On the other hand, freeze-drying is also a method to preserve the activity of these enzymes. The enzymatic activity of the freeze-dried fruits and vegetables do not decrease. The relative concentration of enzymes increases during the freeze-drying process, which will lead to the discoloring, staling, and nutrients loss of freeze-dried foods.

Blanching is necessary before freeze-drying because it can inactivate the enzyme activity, stabilize the color and reduce the loss of Vitamin C. Blanching is also used to exclude the air inside the tissue to avoid oxidation; to soften the tissue and to kill some of micro-organisms attached to the raw materials. Blanching can be done by hot water or by steam. The temperature and time for blanching depend on the types, sizes and blanching techniques of fruits and vegetables.

Slicing is one of the pretreatment processes. It is necessary to slice the fruits and vegetables of big size to meet the consumption habits. Slicing can improve the heat and mass transfer efficiency greatly. In general, the freeze drying time is a function of the volume of the food. The freeze-drying time will increase significantly with the increment of the size of the foods' pieces. Small size of foods' pieces will shorten the freeze drying time because the sublimation area increases. However, the foods should not be sliced too small, else more nutrients will be lost with the drip of juice [6].

2) Pretreament of meat products

The pretreament of meat products includes removing excessive fat, cutting, cooking, or adding the necessary antioxidants.

When the freeze drying time is too long and the temperature is too high, the fat in meat products may melt and result in off-odor and coloring. The melted fat will plug the pores in the freeze-dried layer and block the escape of vapor during the drying process. Therefore, the fat should be removed before freeze-drying. Generally, meat without bone is used for freeze drying because the drying of bones need very long time and the dried bones are useless.

After the fat and bones are removed, the lean meat will be cut into small sliced or diced meat. Cutting surface should be perpendicular to the direction of the muscle fibers in order to accelerate the sublimation of ice. Cutting can be done when the meat is fresh, and cutting can also be performed when the meat is semi-frozen or totally frozen to obtain good rehydration of the freeze dried meat.

In most cases, the freeze-drying meat is consumed directly, so the meat products should be cooked before freeze-driying.

In the pretreatment process, anti-oxidants can be added (such as glutathione, ascorbic acid and tocopherol to prevent fat, protein and pigment from oxidation.

3) Pretreatment of liquid foods

Liquid foods can be classified into natural and extract liquid foods. Natural liquid foods include milk, egg yolk, egg white, and so on. Extract liquid foods include fruit juice, vegetable juice, tea, coffee, and so on. The pretreatment of liquid foods include sterilization, concentration, granulation, adding antioxidant and anti-caking agents.

The methods of sterilization are types of foods dependent, for example, milk can be sterilized by pasteurization.

5.2.2 Freeze-drying

The general procedures of freeze-drying can also be applied to foods, which include three stages: freezing, sublimation drying and desorption drying.

1) Freezing

The quality of freeze-dried products is dependent on the cooling methods, cooling rates and end temperature of the products. Foods can be cooled or frozen by heat conduction with low-temperature solid (such as shelf) or heat convection with low-temperature air or liquid.

According to cooling rates, freezing is classified as quick freezing and slow freezing. Small ice with great numbers will be formed during quick freezing process and the structure of freeze dried products is more denser. Quick freezing has little damage to the cell membrane and protein of foods. The water-holding ability and the flexibility of the rehydrated food are high. But in drying process, there is greater diffusion resistance for the escape of vapor. Big ices will form during slow freezing and the cell membrane and protein are seriously damaged. However, the drying rate is greater than that of quick freezing because big ices result in small diffusion resistance in slow freezing process. Coffee is usually dried after slow freezing to get dark-brown products.

The end temperature of freezing is also an important parameter for freeze-drying process. In recent years, vitrification provides theoretical guidance to freeze-drying technology. As food contains a variety of components, the crystallization of most foods is not completed at a fixed temperature, but in a temperature range. The solution in the foods will become more and more condensed with the crystallization of ice and become maximum concentrated matrix (MCM). The MCM will be vitrified if the temperature is cooled below a specific value, which is called the glass transition temperature ($T_g^{'}$).of the maximum concentrated matrix. In theory, the end freezing temperature should be 5—10K lower than $T_g^{'}$. But this will increase the production cost and should be considered carefully when designing the technical procedures.

The glass transition temperatures of some carbohydrates are given in Table 3.2. The $T_g^{'}$ values of some fruits and vegetables are shown in Table 5.1.

Table 5.1 The $T_g^{'}$ values of some fruits and vegetables

name	$T_g^{'}$ (℃)	name	$T_g^{'}$ (℃)
potato	−12— −16	strawberry	−33— −41
cauliflower	−25	banana	−35
carrot	−25.5	apple	−41

continued

name	T_g' (℃)	name	T_g' (℃)
tomato	−41.5	peach	−36.5
green peas	−27.5	pineapple juice	−37.5
spinach	−17	apple juice	−40.5
sweet corn(blanched)	−9.5	lemon juice	−43
orange juice	−37	white grape juice	−42.5

2) Sublimation drying

Sublimation drying (or primary drying) is the process to remove the free water in food by sublimation. The heat supplied in this process should be equal to that needed for ice in food to sublime. Sublimation drying and determination of the maximum temperature have been analyzed discussed in Chapter three.

3) Desorption drying

Desorption drying (or secondary drying) is the process to remove the bound water in foods. Because of the large adsorption energy of bound water, sufficient heat must be supplied in this process. The technical procedures should be carefully designed to prevent the foods from collapse and coking. The remaining water after desorption drying is controlled between 2% to 5%.

5.2.3 packaging and storage

1) Prevent oxidation and moisture absorption

The major problem for freeze-dried foods is the quality decrement caused by oxidation and moisture absorption.

It is easier for freeze-dried foods to absorb moisture because of the loose porous structure. The freeze-dried slices will contract when absorbing moisture. The freeze-dried granular powders will cake and agglomerate when absorbing moisture. It is necessary to prevent freeze-dried food from moisture absorption because freeze-dried food will lose its intrinsic properties, or even deteriorate after absorbing moisture.

The amount of water contained by foods is fixed at a specific temperature and relative humidity. The capacity of moisture absorption can be determined by the water contents in foods in various relative humidity. Karel reported that the quality of freeze dried foods in storage was affected by water contents and the kind of bond to the solids, that is, the shelf life of freeze dried foods depends on the type of foods. This can be described by the adsorption isotherms as shown in the Fig. 5.1[3].

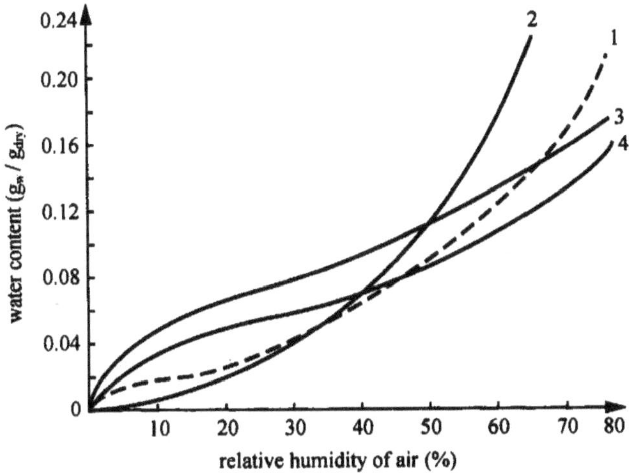

Fig.5.1 Adsorption isotherm of some foods at 22℃[3]

1.Green pepper; 2.peach; 3.potato; 4.beef

The specific surface area of freeze-dried food is large, the contact area with the oxygen in the air is also larger, so The fat and oil-soluble components in freeze-dried food (carotenoids, chlorophyll, ascorbic acid, etc.) will be oxidized when exposed to oxygen, which will results in the discoloring, fading and even deteriorating of foods.

Micro-organisms, enzyme and non-enzymatic reaction can induce degeneration, corruption and browning of foods and drugs. Water activity (a_w) is one of the important factors that affect these reactions. The effect of water activity to these reactions in foods was discussed in Fig. 2.6. The growth rate of bacteria, fungi and yeast will desease with the decrease of the water activity. Maillard reaction rate is not a monotonic function of the water activity. For example, auto-catalytic hydrolysis of fat is still active at very small a_w. Generally, the shelf life should (but do not have to) increase with the decrement of a_w. The optimum water activity of foods should be determined to keep the highest quality of foods.

Kapsalis etc. reported that the residual moisture content of freeze dried peas can not be too big or too small. They found that the thiamine content almost has no change while the carotene content reduces to about 36% of the initial contents when the freeze dried peas (< 5% RM) were stored at 43 ℃ for 84 days. When the RM is between 33% to 81%, the carotene content decreases to 50 % of the initial contents. So the optimum residual moisture content of freeze dried foods is related to many factors, such as the conditions of storage. Medas, Sauvageot and Simatose drew the similar conclusion with the freeze-drying of strawberry and orange juice[3].

The optimum residual moisture content depend on the types of foods and cooling rate. For example, the aroma residuals of foods is related to not only the residual moisture content but also the freezing rate, the thickness of the juice layer, the original concentration, operating pressure, and so on. Kompany and Rene studied the changes of five types of aroma in mushrooms before and after freeze-drying. To keep the maximum contents of these aroma, they suggested that the higher shelf tempreture (90℃) and low chamber pressure (5×10^{-2}mbar) should be used in the initial stages of drying (until about 50% of the water was sublimated), then the temperature was decreased to 60℃ and pressure was increased the 0.5mbar[3].

Lime studied the freeze-drying of the avocado salad (composed of 88.7 % avocado flesh, 4.6% lemon juice, 0.7% onion powder, 1.43% NaCl, and 5.0% crispbread powder). The shelf temperature was 38℃. The final residual moisture content is within 1% to 8%. The freeze-dried avocado salad was canned at 20 ℃ and 30% of relative humidity. The products were canned and sealed in three different ways: vacuum, filling nitrogen and in air. The cans were stored at -18℃, 5℃, 20℃ and 37℃ for 48 weeks. The quality parameters were tested at various intervals from 0 to 48 weeks. The results showed that the freeze-dried avocado salad canned in air degenerated when stored at 20 ℃ for 8 weeks. The shelf life of the freeze-dried avocado salad canned with vacuum or filling nitrogen methods was 16—24 weeks at 20℃ and 48 weeks at 5℃. Lime obtained the relationship between residual water contents and free fat acids (Fig.5.2 A) as well as peroxidase (Fig.5.2 B) by testing 39 samples of canned avocado salad stored at 37℃ for 10 days. It can be seen from Fig.5.2 that free fat acid increases with the increment of residual water content and the peroxide decreases with the increment of residual water content. The optimum residual moisture content is about 2%—3% for the freeze-dried avocado salad.

2) Packaging materials

Generally, packaging material is required to have the following properties:

safe, non-toxic, waterproof, airtight, lightproof, high mechanical strength, suitable for mechanical filling and sealing, easy for transportation.

Brown glass bottles or metal cans should be used if the products are required long-term storage (more than two years). In recent years, the aluminum foil bags partially replaced the metal and glass container. From the aspect of price and weight, it is better to use the aluminum foil bags as packaging materials for freeze-dried food. The aluminum foil bags have a variety of laminating methods, such as, aluminum foil + Polyethylene; cellophane + Polyethylene + aluminum foil + polyethylene; paper + Polyethylene + aluminum foil + Paper + polyethylene; cellophane + Polyethylene + aluminum foil + paper + Polyethylene.

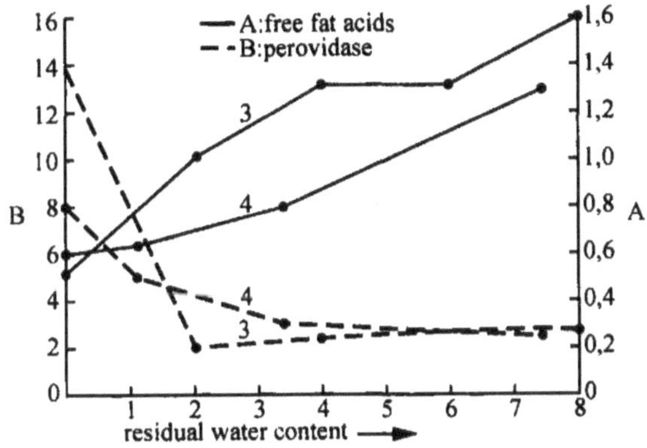

Fig.5.2 The changes of peroxide and free fat acid of freeze-dried avocado salad stored at 37℃ for 10 days [3]

3: sample group one; 4: sample group two; A: free fat acids; B: peroxide

No matter which packaging materials are used, the oxygen in the containers must be replaced with nitrogen. The concentration of residual oxygen should be less than 2%. The desiccants inside the package can prevent moisture absorption of the freeze-dried food. The common desiccants are activated carbon, silica gel and so on.

The packaged freeze-dried food should be stored at the possible low temperature environment. If the storage temperature is too high, the shelf life will be shortened due to the caking, discoloring, collapse.

5.2.4 rehydration

Some of freeze-dried food can be eaten directly; while others need rehydration before consumption.

5.3 Examples of food freeze-drying

The three main processes of food freeze-drying were introduced in section 5.2. The fruits and vegetables, meat, as well liquid food will be taken as examples in this section to illustrate further the techniques of freeze-drying.

5.3.1 Example of fruit freeze-drying

Strawberry is taken as an example to illustrate the technique of fruit freeze-drying [71]. Fesh strawberries of the same size, same types and same maturity were chosen. They are washed in clean water and then dried by natural dripping.

The strawberry's eutectic temperature is determined to be −15℃, therefore the pre-frozen temperature of freezing stage is controlled at −20℃. The freezing rate is around 0.3℃/min which is slow freezing. It takes 2.5 hours to lower the temperature from the room temperature to −20℃ and hold for 0.5h at −20℃.

The ice inside the strawberry starts to sublimate when vacuuming the chamber. The ice front progresses from the surface to the center of the strawberry during the sublimation drying process. The temperature of the frozen parts is kept below −15℃ and that of the dried parts is controlled below 30℃. To remove the bound water in the strawberry, the drying temperature is increased to the limit of 40℃ because the Vc in the strawberry will decompose when the temperature is greater than 40℃. The freeze drying process (~4h) is finished when the vacuum is very close to the limit vacuum. The freeze dried strawberry is packaged in hard plastic boxes filled with nitrogen and stored at room temperature.

5.3.2 Example of vegetables freeze-drying

The carrots are taken as an example to introduce the freeze-drying techniques of vegetables [8]. Carrots should be stored at the temperature close to 0℃ and the relative humidity not lower than 95%. The raw carrot should be big size and maturity containing high solids, high carotene and fewer fibers. The raw carrot should not be rotten, polluted, blighted by insects, and mechanically damaged. The freeze drying techniques is shown in Fig.5.3.

Classification→ raising→ peeling→ washing→ trimming→ check→ slicing→ washing→ blanching→ sulfite treatment→ freezing→ drying→ selection → checking→ packaging→ boxing→ storage→ delivery

Fig.5.3 The freeze drying techniques of carrots

First, the carrots were dried in the tunnel sieve to remove the dust, remnant leaf and small pieces of carrots. Then, the skin of the carrots was removed by steam or lye methods. When the steam method was used, the carrots should be processed under 0.68MPa for 30 seconds. The processing time should be prolonged if the processing pressure is lower than 0.68Mpa. When the lye method was used, the carrots have to be washed by 5% lye at 99℃ for 4 minutes. The carrots processed by steam or lye methods were sprayed by high pressure water torch to remove the softened skins and remaining lye. After peeling, the carrots were classified into the thick and thin groups by the automatic selector. The carrots' blue head were cut off by the automatic cutter. Then the carrots were checked in the inspecting belt to trim the carrots to remove the rotten parts. After checking, the thick carrots were cut into dices and the thin carrots were cut into round slices. The carrot pieces were tossed onto the moving stainless steel mesh in a distribution of 18kg/m^2 and heated by steam for 6—8min. At the end of the mesh, the carrots slices were sprayed with 0.9%—1.0% sulfite solution (The concentration of the sulfite solution was determined to guarantee the SO_2 content is 500—1000mg per kg carrots). Sometimes the 2.5% corn starch at 79℃ was used to spray the blanched carrots slices.

Carrots contain high sugar contents, so the freeze drying temperature should be kept below −33℃. If the carrots were frozen by the reduce pressure method, the pressure of the freeze drier should be kept at 27Pa. The carrots must be frozen outside the drier if the pressure of the drier cannot be kept below 27Pa.

The freeze dried carrot slices were packaged by paper - aluminum foil - polyethylene compound materials and stored at room temperature.

5.3.3 Freeze-drying examples of liquid foods[8,72]

Instant coffee is taken as an example to inllustrate the freeze-drying of liquid foods. This is a typical application of freeze drying in food industry. The liquid coffee should be concentrated before freeze drying to reduce products cost. The solids in the concentrates should be greater than 40%.

The technical procedures of freeze-drying for instant coffee include: baking coffee bean, extraction, concentration of extracted liquid, freezing concentrated solution, sublimation drying, desorption drying, smash, and package. The parameters to detect the quality of instant coffee include: color, package size, uniformity of pellet granularity, flavor and so on. These parameters are affected by the quality of raw materials, concentration methods and the technical procedures of freezing/drying.

Compared to drying process, the freezing has greater effect on the color, flavor and density of freeze dried coffee. The effects of cooling rate and cocentration of extracted solution on the quality of freeze dried coffee are shown in Table 5.2.

Table 5.2 The effects of cooling rate and cocentration of extracted solution on the quality of freeze dried coffee

freezing time (min) (from 0℃ to − 15℃)	cocentration of extracted solution(w/w%)	drying time(h)	bulk Density g/(100mL)	color	flavor
0.5	24	9.0	10.8	lightest	best
7	24	10.5	11.6	lighter	better
15	24	11.5	13.5	light	better
26	20	13	15.3	normal	good
27	20	12.5	16.2	normal	good
28	30	12.0	16.7	normal	good
60	24	13	16.8	darker	acceptable
67	14	14.5	18.4	darkest	acceptable
120	30	16.0	25.0	blacker	worst
210	24	13.0	27.3	blackest	worst

It can be seen from Table 5.2 that the color of the freeze dried products become more lighter with the increase of freezing rate. The concentration has no influence on color when the concentration of soluble solids is between 20% to 30%. For extracted solutions with equal concentration, drying rate increases with the increment of freezing rate. The density of products reduces with the increment of freezing rate. The flavor of freeze dried coffee increases with the increment of freezing rate.

The black coffee pellets were obtained by slow freezing, while lighter color coffee pellets were obtained by fast freezing. Many consumers usually think that the sepia coffee has higher quality. The color becomes darker with the increment of drying pressure. High grade, sepia coffee can be produced by increasing the drying pressure to a specific value.

The relationship between the size of coffee pellets and the bulk density is shown in Fig. 5.4. The concentrated solution pretreated with CO_2 was frozen at −40℃ for 20min. by using the plate freezer. The bulk density decreased by 25% after freezing process, but the color does not change.

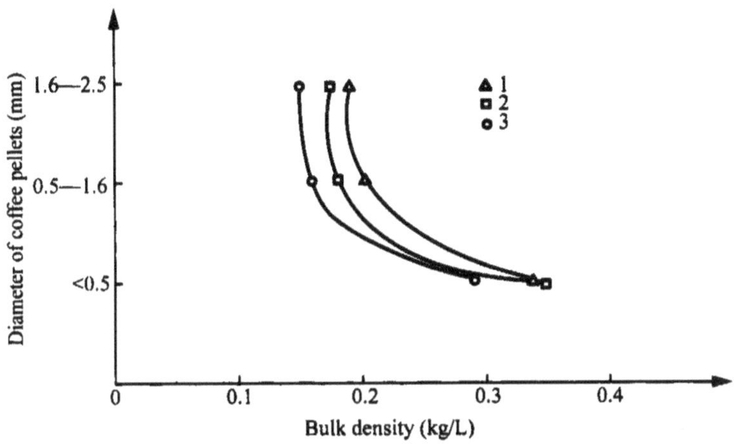

Fig. 5.4 The relationship between the size of pellets and bulk density of coffee

Pretreated with CO_2 under different pressure. 1: No CO_2; 2:2bar; 3:12bar

Flink gives the relationship between the relative retention rates of volatile materials in concentrates and freezing rate/pressure as shown in Table 5.3.

Table 5.3 The relationship between the relative retention rates of volatile materials in concentrates and freezing rate and pretreatment

Freezing methods	Pressure in the chamber(mbar)						
	0.26	0.4	0.53	0.66	0.8	0.92	1.06
	relative retention rates(%)						
Slow freezing[1)]	92	96	78	77	66	67	34
Slow freezing[2)]	100	99	88	82	91	82	35
Foaming slow freezing[3)]	67	61	49	53	57	44	63
Fast freezing[4)]	47	53	38	38	44	35	36
Foaming fast freezing[5)]	48	—	42	42	43	32	29

1)The temperature of materials (with thickness of 15mm) is decreased to −10℃, −25℃, −40℃, respectively. The time required is about 24h.

2) The material with thickness of 15mm is put on the aluminum tray. Its temperature is dropped to -40 ℃ in still air.

3) Concentrated coffee is foamed in the softing machine, and then the foaming layer with 15mm thickness is frozen to −40℃ in still air.

4) Coffee concentrates were sprayed on the drum surface at −52℃,and frozen layer is fromed.in 3mm.

5) Coffee concentrates pretreated with CO_2 is foamed at −20℃, and then sprayed on the drum surface at −52℃

After frozen concentrates is grinded, the coffee pellets with size of 1.2~2.7mm fill the aluminum tray with the thickness of 15mm. The tray is put into the space between two shelves for drying. The temperature of shelf is maintained below 40℃. The highest relative retention rates of volatile material that can be obtained by slow freezing under 0.26mbar are shown in Table 5.3.

The technical procedures for freeze-drying of coffee is shown in Fig.5.5 and the production procedures are shown in Fig.5.6.

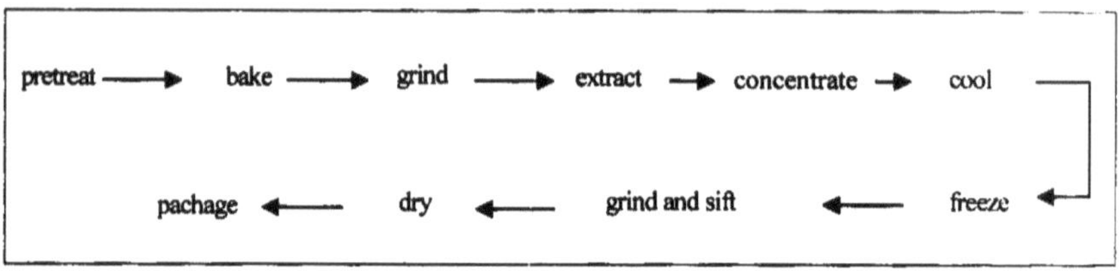

Fig. 5.5 The technical procedures for freeze-drying of coffee

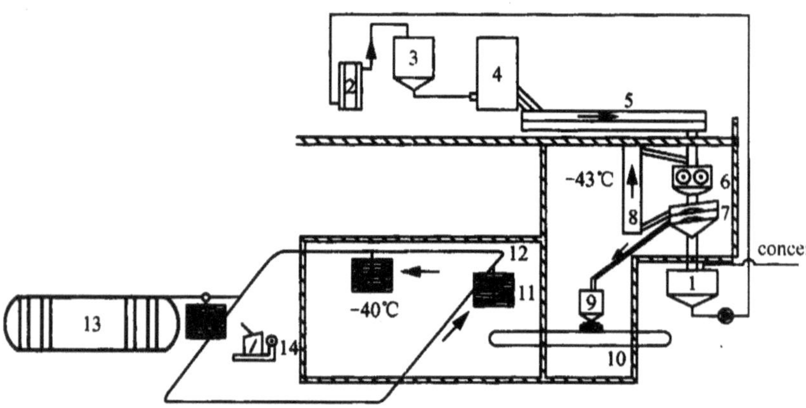

Fig. 5.6 The production procedures for freeze-drying of coffee

1. tank for concentrates; 2. heat exchanger; 3. tank; 4. ice slicing machine; 5. freezing tunnel; 6. twin-roll grinder; 7. two stage vibration sieve; 8. lifter; 9. tank; 10. transport belt; 11. crane; 12. track; 13. freeze dryer chamber; 14. balance

The concentration of raw coffee concentrates for freeze-drying is 40%. The purpose of concentration is to increase the drying efficiency and control the density of the freeze dried coffee pellett. The initial temperature of concentrated solution is 50℃. This solution should be cooled to 0—3℃ to reduce the valotile of aromatic substances, plate-type heat exchanger(2) is usually used to cool the solution.

The cooled concentrates solution was piped to ice slicing machine(4). The solution was frozen in the ice slicing machine. The vaporizing temperature of the refrigerant is –30℃. The thickness of sliced frozen concentrates is 5—10mm and the temperature is about –10℃.

Finer coffee particles are desired to increase drying rate and fullfill the habit of consumers. But the temperature of the materials from ice slicing machine(4) is only –10℃, the materials at –10℃ is not hard enough for grinding. And the heat In order to make the material more elevated area in order to improve the drying rate; at the same time as generated by grinding will result in partially melting of the materials. So the materials should be cooled to –43℃ by a freeze tunnel (5) before grinding by twin-roll grinder(6).

The temperature of the freezing storage room is about –40℃. In the storage room, trays of materials are mounted on a small crane (11). The materials were sent to freeze drying chamber (13) by the small cranes (12), rolling through the track. Close the door and vacuum the drying chamber to 0.5mbar. The coffee particles were dried at –27℃ under 0.5mbar. The heat needed for sublimation is provided by the heater inside the chamber. The sublimated vapor was condensed on the cold trap.

The freeze dried coffee particles should be packaged into thin film bags immeditely to prevent moisture absorption. The package room should keep the temperature of 25 ℃ and relative humidity below 30%—40%.

5.3.4 Examples of meat freeze drying

The freeze drying technical procedures of meat are introduced taking sliced beef as an example.

Generally, the freeze-drying procedures for raw beef include: Cooling the carcass to 3℃ after slaughtering, separate and store (for aging) for one week in low-temperature bag; freezing by air-freezer at –30℃ and frozen storage for 1—3 days; slicing; vacuum drying. In the drying process, the highest temperature for heating shelves, materials surface and ice sublimation interface should be kept at 120 ℃, 40 ℃ and –25 ℃, respectively. The dry process ends when the moisture content is between 1% to 1.5%. Finally the freeze dried beef slices were packaged in nitregon (<0.1% oxygen).

To get the freeze dried beef with good rehydration, the thickness of the beef sliced should be less than 1cm, and slicing is done for the frozen or semi-frozen beef. The cutting direction should vertival to the direction of the fibers. The bone, fat and connective tissue should be elimited.

The freeze rate will affect the drying rate, the porous and rehydration of freeze dried beef. The effect of freezing rate on the quality of freeze dried beef is shown in Table 5.4.

Table 5.4 The effect of freezing rate on the quality of freeze dried beef

quality	freezing rate/(cm/h) (the progresssing rate of the frozen front)		
	0.2 1)	0.6 2)	20—30 3)
dripping(9 grades)	4.8	4.3	4.2
rehydration/%	92	92	81
brightness (Hunter CDM)	28	29	41

1) freezing in −12℃ air; 2) freezing in −30℃ air;3) freezing by nitregon spraying.

It can be shown from Table 5.4 that too fast freezing rate will decrease greatly the rehydration capacity and induce more light color.

In order to prevent the melting of frozen beef, the temperature of frozen beef cannot be higher than −25℃ for the beef with fat content of 5%. The temperature of the dried layer of meat should be at 20—60℃.

5.4 Hazard analysis and critical control point for freeze-drying of food

5.4.1 Introduction of HACCP[73—75]

1) Definition of HACCP

HACCP, the acronym of "hazard analysis and critical control point", is a simple and science-based food safety system for assuring the safety of food from chemical, physical and biological hazards from the farm to the table.

HACCP is different from traditional method of quality control in that the latter is the control of testing for failure (end-product testing); while the HACCP is a preventive system. HACCP is applied at each food processing step for assuring safe production of foods, including processing of raw material, storage, transportation and consumption. In recent years, HACCP has been recognized as a new safety management system in food industry by most of the countries of the world.

2) Origins and development of HACCP

The application of the concept of Hazard Analysis and Critical Control Points (HACCP) for safe food production was pioneered in 1960s by the Pillsbury Company, US Army and US National Aeronautic and Space Administration

(NASA) as a collaborative development for the production of safe food for the US space program. There were two safety issues in the space capsule under zero gravity conditions, 1) crumbs and water droplets getting into electrical equipment, 2) preventing foodborne illness which would have catastrophic results in a zero gravity environment. The research program looked into ways of obtaining as close to 100% assurance as possible that food products for astronauts would not be contaminated with microbiological, chemical, or physical hazards. The Pillsbury Company, therefore, introduced and adopted HACCP as the system that could provide the highest level of safety, whilst reducing dependence on inspection and end-product testing. In the early 1970s, Pillsbury transferred the HACCP concepts from the space program to production of its commercial food plants. Pillsbury presented the HACCP concept at the National Conference on Food Protection in 1971. In 1972, The HACCP concept was carefully discussed at the National Conference on Food Protection and was transferred to the FDA for training FDA personnel in the same year.

The use of HACCP principles in the promulgation of regulations for low-acid canned food was completed in 1974 by the US Food and Drug Administration. In 1977, the HACCP concept was used in seafood for the first time. US Academy of Science endorsed the broad application of HACCP in food safety in 1985.

In 1989, NACMCF(National Advisory Committee on Microbiology Criteria for Foods) drafted "The Codex Guidelines for the Application of the Hazard Analysis Critical Control Point System in Food". In 1993, the Food Hygiene Committee of the Codex Alimentation Commission drafted "the guidelines for the application of HACCP principles", and regulated the seven basic principles of HACCP (being implemented all over the world nowadays), such as terminologies, development, and the application of CCP diagram.

In 1997, FAO (Food and Agriculture Organization) and WHO (World Health Organization) extended food inspection to the whole process of food production, including raw-material production, storage, transportation and consumption. The Codex Alimentations Commission has promulgated "the application guidelines of HACCP system".

Since the 1990s, Food Branch of Science and Technology Committee, Administration of Import and Export Commodity Inspection of China, has begun to study the application of HACCP in food processing industry and issued the guidelines of HACCP system in export food production and food processing. Institute of Food Inspection, Ministry of Public Health, has begun to study HACCP management system in the production of dairy product, cooked meat and beverage since 1991. In 1999, Ministry of Agriculture issued the Industry Standard of Seafood—the quality management standard of seafood SC/T3009—1999 to enforce the application of HACCP. State Administration Committee of Certification and Accreditation released "Certification regulations on hazard analysis and critical control point (HACCP) management system of food industry in China" in 2002.

3) Advantages of HACCP

HACCP has the following advantages as a new safety assurance system:

(1) HACCP is a systemic and preventive system for food safety compared with the traditional end-product testing method.

(2) HACCP is a management tool that can be used to manage the risks associated with food at all stages of the food chain, from farm to table. It is an active method to prevent the foodborn problem.

(3) HACCP system concentrates the critical point of production process to ensure the quality of product.

(4) HACCP system control the product quality by monitoring temperature, time and appearance, etc., which is simple, directive, fast and feasible.

(5) HACCP system requires all the stuff in the company, including technician and administrator, to be involved in the management of food safety.

(6) HACCP does not mean zero risk because it cannot ensure to eliminate the potential hazards completely.

5.4.2 Typical hazards in food production

Hazards in food production are defined as the factors that will cause safety problems to the consumers. Based on their origin, these factors are classified in microbiological, chemical and physical hazards, which account for 93%, 4% and 3% in epidemic events, respectively. It can be seen that the microbiological hazard is the major source of foodborn illness. However, chemical hazard can also lead to foodborn diseases, and physical hazard will affect food safety, too.

1. Microbiological hazards

The microbiological hazards in food production are defined as the contamination of food raw materials, processing and end products caused by organism (especially micro-organism) and its metabolic products (such as toxins). The microbiological hazards are classified as bacterial hazard, fungal hazard, virus and Rickettsia, parasitic disease, and insects hazard, etc.

2. Chemical hazards

Chemical hazards in food production are defined as the contamination of food caused by toxic chemicals. The main sources for chemicals are: natural toxins, pesticide residues, veterinary drugs residues, heavy metal ion, abused food additives, food packing material, containers and equipment, and radioactive contamination.

3. Physical hazards

Physical hazards in food production are defined as any abnormal physical materials found in food which is pathogenic to human beings. The main sources are: contaminated material, facilities and equipment with poor design and maintain, mishandling in processing.

5.4.3 Five preliminary procedures and seven principles in HACCP system

1. Five preliminary procedures

(1) Collection of general information.
(2) description of the food.
(3) description of food distribution and storage methods.
(4) Identification of the purpose and consumption of food.
(5) development of a flow diagram.

2. Seven principles

1) Conduct a hazard analysis

Identify all potential hazards in food production at all stages, including raw material processing, food manufacture, products storage and transportation, consumption, etc. Assess the likelihood of occurrence of hazards and identify the measures for their control.

2) Identify the critical control points (CCPs)

The critical control point (CCP) means the step or steps at which control is essential to prevent or eliminate a hazard or to reduce it to acceptable levels. Only when the points being regarded as significant food security hazards were controlled, they become the critical control point. The identification of critical control point is usually based on a decision tree (Fig.5.7).

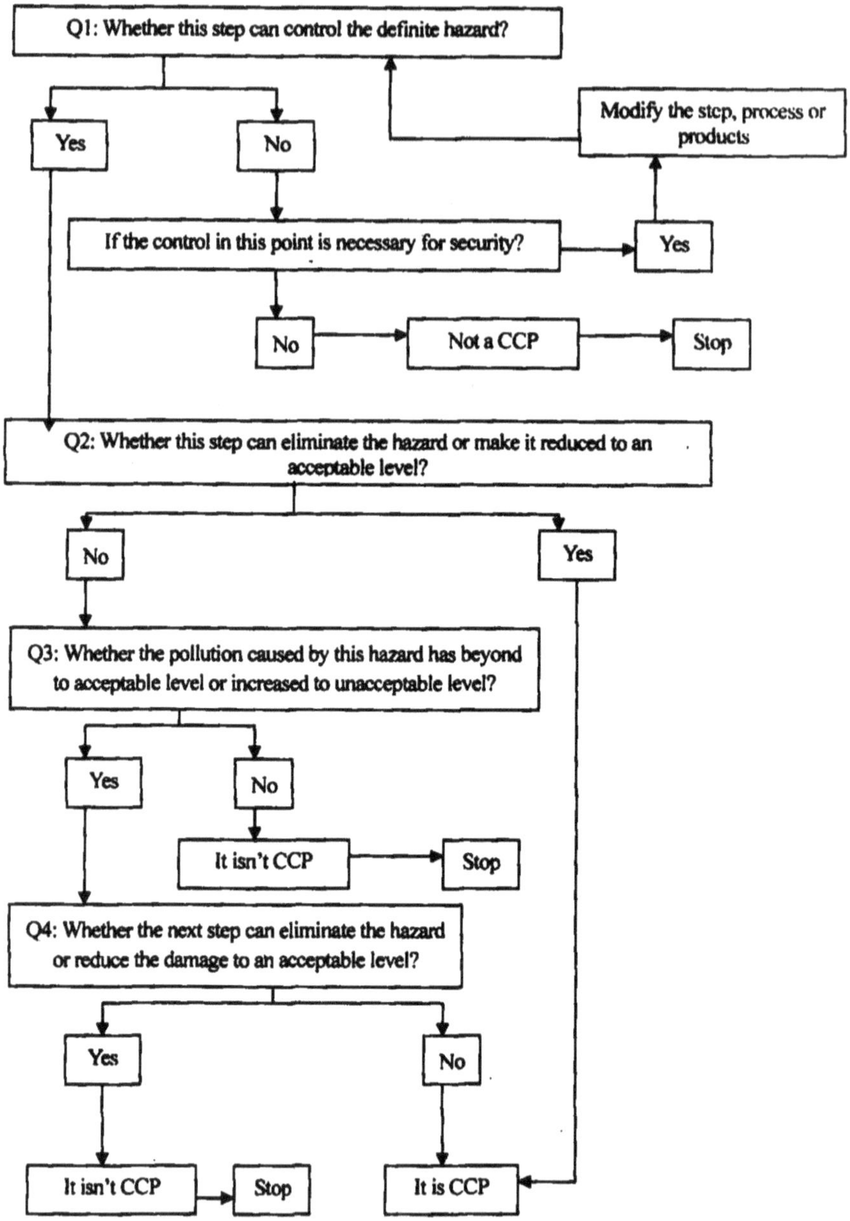

Fig. 5.7 Decision tree for critical control points

3) Establish Critical Limits

The critical limit is a maximum and/or minimum CCP value to which a biological, chemical, or physical parameter must be controlled to prevent or eliminate food safety hazard, or to reduce it to an acceptable level. A standard value is needed for each CCP to ensure the CCP was characterized as safe. These critical limits are usually parameters related to food safety, such as temperature, time, physical properties (such as tension), moisture content, water activity, PH value.

4) Establish monitoring procedures

Monitoring is scheduled testing or observations to assess whether a CCP is under control and to produce an accurate record for future evaluation. The CCP should be monitored on-site through a variety of physical and chemical methods. Else, Observation of CCP changes must be very frequently to ensure the CCP is under control.

5) Establish corrective actions

Establish effective procedures for corrective actions when monitoring indicates that a particular CCP is not under control. Although there are plans in a HACCP system to prevent from deviations, each of the CCP should have the appropriate corrective schemes to recover or correct the deviations and record the corrective actions in case of deviations.

6) Establish documentation and record-keeping procedure

Establish appropriate documentation of the HACCP plan. It is required to record completely the information and data on implementation of a HACCP plan, including preparation, execution, monitoring and records of a HACCP plan.

7) Establish verification procedures

Verification procedures are used to confirm the proper implementation of the HACCP plan Verification should be recorded to show the procedures of implementation at any point.

5.4.4 The relationship between HACCP and other quality assurance systems

1) The relationship between HACCP and ISO 9000

ISO 9000 series standards (including ISO 9000, ISO 9001, ISO 9004, ISO 9011, etc.) are universal international quality management systems, issued by Quality Management and Quality Guarantee Technical Committee of International Organization for Standardization (ISO/TC176). ISO 9000 series standards are issued to meet the requirements of customers and the law, while HACCP is used to control the food hazards. There are many similarities between them, for instance, the well-structured management system and participation of all staff. The difference between them is that ISO 9000 series standards are standard, comprehensive and recommended, but HACCP is principle, mandatory, and targeted.

ISO 9000 can be applied to the quality control system of all industries. HACCP is based on the seven principles and included in mandatory regulations of food processing in many countries. It is applied to a unique food system that requires safety control. HACCP and ISO 9000 cannot be substituted for each other even they are both used to ensure the quality and safety of products. It is a tendency that combining HACCP and ISO9000 together to assure food safety and quality.

2) The relationship between HACCP and GMP

Good Manufacturing Practice (GMP) is a professional quality assurance (QA) or manufacture management system. GMP is mostly applied to pharmaceutical industry. Many countries also establish food GMP regulations for food production management. Many GMP regulations have been issued in China, such as "Good Manufacturing Practice (GMP) for functional food" (National Standard GB17405—1998) approved by Chinese Ministry of Public Health in 1998. GMP focuses on the control of food processing (including production environment) and storage/distribution. GMP is the base of implementation of HACCP.

3) The relationship between HACCP and SSOP

Sanitation Standard Operating Procedure (SSOP) is generally documented steps that must be followed to ensure adequate cleaning of product contact and non-product surfaces. These cleaning procedures must be detailed enough to make certain that adulteration of product will not occur. All HACCP plans require SSOP to be documented and reviewed periodically to incorporate changes to the physical plant. This reviewing procedure can take on many forms, from annual formal reviews to random reviews, but any review should be done by "responsible educated management". As these procedures can make their way into the public record if there are serious failures, they might be looked at as public documents because they are required by the government. SSOP in conjunction with the Master Sanitation Schedule and Pre-Operational Inspection Program, form the entire Sanitation operational guidelines for food related processing and one of the primary backbones of all food industry HACCP plans.

5.5 Application of HACCP in food freeze-drying

This part will take kiwi fruit as an example to describe the application of HACCP in food freeze drying[76].

5.5.1 Information and description of product and procedure

1. Description of product

Freeze dried kiwi fruit is a type of ready-to-eat product that was made from fresh kiwi fruit through a series procedures as washing, sterilizing, peeling, slicing, quick freezing, freeze drying and packaging.

Freeze dried kiwi fruit contains no preservatives and additives. It has the character of bright-color, uniform size, nutritional and good rehydration property. The loss of aroma fragrance is relatively low during lyophillization because its volatility is lower at low temperature. In addition, the concentration of aroma increases after drying. As a result, the flavor of freeze dried kiwi fruit is kept intrinsic. Freeze-dried kiwi fruit is distributed and sold at room temperature.

2. Technical procedures (Fig. 5.8)

The technical procedures for freeze drying of kiwi fruits are shown in Fig. 5.8.

(1) Refrigeration: It is recommended that fresh kiwi fruit is stored in cold warehouse at (0±0.5) ℃, with 95% relative humidity and higher CO_2 concentration to keep the fruits' hardness and freshness.

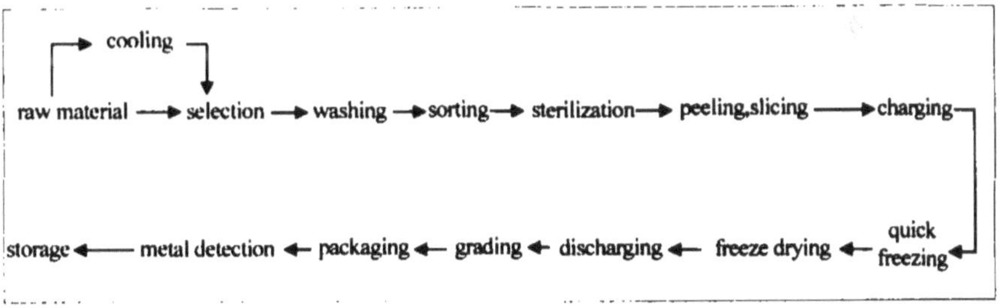

Fig.5.8 Technical procedures for freeze dried kiwi fruits

(2) Selection: Special equipments are applied to determine the soluble solid content, fragrance content and sugar/acid ratio in order to eliminate the immature fruit whose flavor is below the mark. At the same time the tender fruits were selected and eliminate the mechanical-damaged, pest, rotten,

overmature and abnormal fruits. It is also desired to eliminate the fruits whose diameter is smaller than 30 mm. Selection is done in the pretreatment room.

(3) Washing: The kiwi fruits are sent into the washer through a conveyer belt and cleaned with the high pressure water gun. The pressure of the water gun should be controlled because the dirt adhering to the fruits' surface cannot be cleaned thoroughly under too low pressure, while physical damage of the surface and sequentially the microorganism contamination will be caused by too high pressure. Washing is performed in the pretreatment room.

(4) Sorting: The washed fruits are sorted by size to ensure the uniformity of sliced fruits. Physical damage of the fruits must be avoided during sorting process. Sorting is performed in the finishing treatment room.

(5) Sterilization: The kiwi fruits are immersed in the sodium hypochlorite solution to kill the pathogenic bacteria on the surface. The concentration and immersion time are determined by the industry standards. Spray the fruits with clean running water (not stagnant water) after sterilization to remove all of the chemical residues. Sterilization is performed in the postprocessing room.

(6) Peeling: Sterilized kiwi fruits are cooled at –25℃ for 30 min to freeze the skin and then washed by high pressure water to make it convenient to peel and slice. The fruits should be sliced right after peeling to avoid the Vc oxidation. Sterilization is operated in the low temperature room.

(7) Slicing: The fruits are sliced by using microtome with the slice thickness of 3—4 mm. Slicing should be done as fast as possible to avoid browning and Vc oxidation. Slicing at low temperature can reduce the loss of the nutrients and Vc. Slicing is operated in the low temperature room.

(8) Charging: The slices must be put rapidly on the tray in the style that can increase the surface area of sublimation, shorten freeze-drying time. Charging is operated in the charging area of the low temperature room.

(9) Quick-freezing: The freezing rate will affect the porosity, rehydration, nutrient contents and drying rate. Fruit and vegetable are generally frozen at – 35— –40℃ in the quick-freezing room.

(10) Vacuum drying: The frozen slices are dried in the drying room with the vacuum of 110—133Pa and shelf temperature of 90—100℃.

(11) Discharging: The dried slices are unloaded from the trays and put into the big plastic bags in the discharging room.

(12) Grading: In the grading room, the dried slices are graded by size. The defective slices are eliminated.

(13) Packaging: The dried slices are packaged into the composite plastic film by vacuum nitregon-filling package method.

(14) Metal detection: The products are detected to exclude the metal fragment or glass that may mix with the products during the production process.

(15) Storage: The products are packed into the boxes and then stored in the products room.

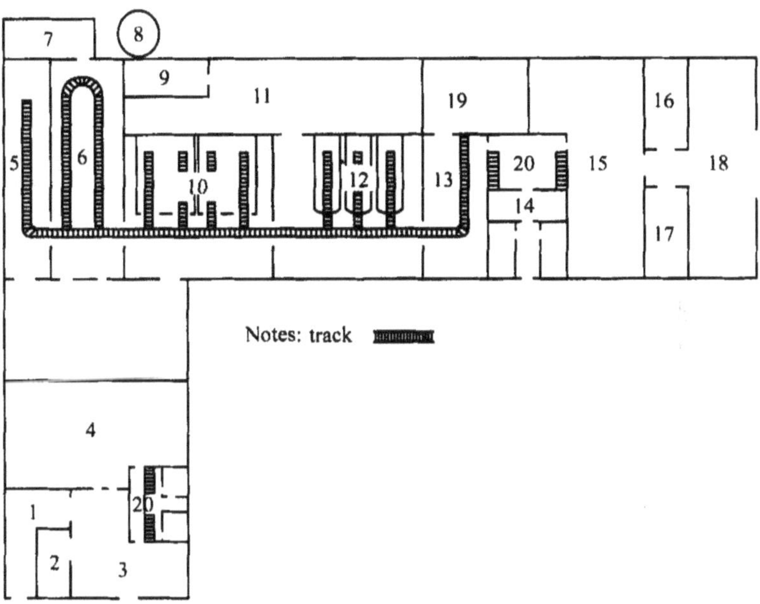

Fig.5.9 The technical procedure for freeze drying of kiwi fruits

1,2. cold warehouse; 3. pretreatment room; 4. treatment room; 5. low temperature room; 6. charging area; 7. boiler; 8. cooling tower; 9. controlling room; 10. quick freezing room; 11. mashine room; 12. drying room; 13. discharging room; 14. spraying room; 15. selection room; 16. packaging room; 17. metal detection room; 18. products room, 19. transition room; 20. worker passage aisle

Fig.5.9 shows the technical procedures for freeze drying of kiwi fruits. The raw material passages and the worker passages must be separated.

5.5.2 Hazard analysis

There are three hazard factors in the production of freeze-dried kiwi fruits, which are microbiological contamination (bacterium, parasite, mold and so on); chemical pollutants (pesticides, heavy metals and detergents) and physical pollutants (foreign objects). These factors involved in the technical procedures are analyzed in Table 5.5.

Table 5.5 Hazards analysis of kiwi fruits freeze-drying—identification of critical control points

(1) procedures	(2) potential hazard	(3) significant	(4) significance for judgment	(5) methods to prevent hazards	(6) CCPs
raw material	B: bacteria C: residues (pesticide) P: foreign subjects	N Y N	high frequency of chemical hazards; affect the product sanitary quality; other potential hazards can be controlled by GMP and SSOP	certification of inspection, sensory test and examination of chemical/physical properties	Y
refrigeration	B: bacteria C: no P: foreign subjects	N N	controlled by GMP and SSOP controlled by GMP and SSOP		N
selection	B: bacteria C: no P: no	N	controlled by GMP and SSOP		N
washing	B: bacteria C: pollution P: exceeding water pressure	N N N	controlled by GMP and SSOP controlled by GMP and SSOP controlled by GMP		N
sorting	B: bacteria C: no P: no	N	controlled by GMP and SSOP		N
sterilization	B: bacteria C: sanitizer residues P: no	Y Y	bacteria and sanitizer residues remained in this procedure will affect the product sanitary quality	washing by high pressure spray with required sanitizer concentration and sterilization time	Y
peeling & slicing	B: bacteria C: no P: metal pieces	N Y	controlled by GMP and SSOP affecting the safety	eliminated by metal detection	N

continued

(1) procedures	(2) potential hazard	(3) significant	(4) significance for judgment	(5) methods to prevent hazards	(6) CCPs
charging	B: bacteria C: no P: no	N	controlled by SSOP		N
quick-freezing	B: bacteria C: no P: no	N	controlled by GMP and SSOP		N
vacuum drying	B: bacteria C: no P: no	N	controlled by GMP and SSOP		N
discharging	B: bacteria C: no P: foreign subjects	N N	controlled by GMP and SSOP controlled by GMP and SSOP		N
grading & packaging	B: bacteria, spoilage C: toxicity of packing material P: foreign subjects	Y N Y	the temperature and humidity of package and operating room do not meet the standards;	regularly sterilize grading tools; package bags meet sanitary requirements; control the temperature and humidity in operating room	Y
metal detection	B: no C: no P: metal	 Y	remaining metal will affect the food safety.	inspecting the end-products by metal detector	Y
storage	B: bacteria, insects, rats C: no P: no	N	controlled by GMP and SSOP		N

B:biology; C: Chemical; P: physical; Y;Yes; N: No.

5.5.3 Establishment, control and adjustment of CCPs

According to the decision tree (Fig 5.7), the CCPs of freeze-dried kiwi fruits are as follow:

(1) Raw materials: Kiwi fruits may be polluted in growth, particularly the pesticide residues and heavy metals in the soil. Therefore raw materials is one of the major CCPs.

(2) Sterilization: Because of the concentration of solution and sterilization time being not controlled properly, the general bacterial population of the sterilized products will be higher than the standard number. Sterilization solution may remain on the surface of fruit kiwi if the washing is not operated properly.

(3) Grading and Packaging: The bacterial number in packaged products may increases because the air, staff and packaging material may be somewhat polluted. It also related with the operating temperature and packaging time.

(4) Metal inspection: Metal pieces in the production are harmful to human beings.

In order to establish an effective HACCP system for freeze drying of fruit kiwi, the CCPs and its limitation, inspection system and corrective plan must be set up after the analysis of hazard factors. The details are listed in Table 5.6.

Table 5.6 HACCP method of kiwi fruits freeze-drying

identify CCP	establish critical limits	CCP monitoring				corrective actions	record	verification
		object	method	frequency	person in charge			
raw material pesticide residues	BBH≤0.2mg/kg DDT≤0.1 mg/kg Hg≤0.02 mg/kg	sensory test, property test and pesticide residue data	inspection of data and visual inspection of raw material	each batch	quality monitoring department, Inspector	stop purchase; reject defective raw material	raw material inspection, purchase and rejection record	check record everyday; recheck corrective actions record every week
Sterilization	4%—6% sodium hypochlorite, diluted by 500 times; sterilization time >5 min	concentration of sterilization solution and sterilization time	reagent paper for chlorine and stopwatch	per duty	quality controller	resterilize with standard sterile solution, wash with drink water	the concentration of sterile solution, washing time, microorganism test result	Check record and test results everyday
metal detection	products without metal fragment	metal in end products	metal detector, magnetic field	every 4 hours	quality controller, maintenance manager	stop production, adjust equipments, destroy defective goods	inspect whether metal detector works regularly, equipment test record	Check metal extracter everyday; summarize Corrective actions; record everyweek
classification & package	room temper-ature is 25℃ relative humidity is 25%—30%; indoor air is positive pressure; completely sealed package	air temperature, relative humidity, sealing test for pressured package	themometer, hygrometer, pressure gauge, test sealing in length and breadth	test sealing every 15min per duty;	quality controller	stop production if temperature,humi dity, pressure and sealing test do not meet requirements	temperature,humidity and pressure records, sealing test results	Check equipment test record everyday, Check machine situation regularly

6

Protective Agents and Additives for Freeze-drying of Pharmaceutical Products

The function and application of protective agents and additives were introduced in section 1.1.3. In this chapter the protective mechanism of protective agents and additives will be analyzed.

6.1 Sugars/polyols-type protective agents

6.1.1 The definition of sugars and polyols

Sugar is made of carbon, hydrogen and oxygen. Sugar is also named as carbohydrate because the proportion of hydrogen and oxygen inside sugar is always 2:1 which is the same as water. Sugar is classified as monosaccharides, oligosaccharide and polysaccharide. Monosaccharide, with the smallest molecule, is the compound which cannot be further hydrolyzed. The typical types of monosaccharide include glucose, fructose, galactose and ribose. Oligosaccharide can be hydrolyzed into 2 to 10 monosaccharides molecules. Sucrose, maltose, monohydrate, lactose, trehalose dihydrate and raffinose pentahydrate belong to oligosaccharide. Polysaccharide can be hydrolyzed into monosaccharides and oligosaccharides. Starches, cellulose and pectin are polysaccharide.

The alcohol containing two or more hydroxyls is called polyol or sugar alcohol. Glycerol, sorbitol and mannitol are polyols widely used in the freeze-drying of pharmaceutical products. Sugar and polyol have similar protective mechanisms because they have the same functional groups (hydroxyls).

6.1.2 The function of sugars/polyols in freeze-drying

Sugars/polyols are used as protective agents in freezing and freeze-drying of biological products. The common sugars/polyols-type protective agents are shown in Table 6.1. Polysaccharide has the same protective mechanism as that of polymer protective agents because it is the compound of monosaccharides.

Table 6.1 The sugars/polyols used in freeze-drying and storage process[2–4]

type	name	chemical formula	relative molecular weight	glas transition temperature T_g' (℃)	collapse tempera-ture T_{col} (℃)
monos accharide	glucose	$C_6H_{12}O_6$	180.16	-43	-41
	galactose			-41	
	mannose			-41	
	fructose			-42	
	ribose	$C_5H_{10}O_5$	150.13	-47	
	xylose			-48	
oligos accharide	sucrose	$C_{12}H_{22}O_{11}$	342.30	-32	-31
	lactose				30.5
	maltose monohydrate	$C_{12}H_{22}O_{11} \cdot H_2O$	360.32	-30	
	trehalose dihydrate	$C_{12}H_{22}O_{11} \cdot 2H_2O$	378.34		-29
	raffinose pentahydrate	$C_{18}H_{32}O_{16} \cdot 5H_2O$	594.53		-26
polysacc haride	mannitol	$C_6H_{14}O_6$	182.17	-27	
	glycerol	$C_3H_8O_3$	92.09	-65	
	sorbitol	$C_6H_{14}O_6$	182.17	-44	-54
	xylitol	$C_5H_{12}O_5$	152.15	-47	
	inositol	$C_6H_{12}O_6$	180.16		

1. Monosaccharides

Theoretically, Monosaccharides can protect the biological products if the hydrogen bonding, which formed between the molecules of saccharides and the active components of biological products, replace the position of water molecules. However, experiment results show that monosaccharides (such as glucose、 galacrose) cannot protect protein during the freeze-drying process, because the weak stabilization provided by monosaccharide in the freezing step will result in the irreversible denaturation of protein before the drying begin. But trehaloses can provide effective protection during both freezing and drying process. Generally the monosaccharides are usually mixed with other excipients as the protective agents for the freeze-drying of biological products.

2. Oligosaccharides

Oligosaccharides, especially disaccharides, are usually used as protective agents because the disaccharide can be used as both the cryoprotective agents in freezing and the protective agents in drying process [77–80]. Disaccharide can be classified as reducing disaccharide (lactose, maltose) and non-reducing disaccharide (trehalose, sucrose). However, the disaccharide will cause the Maillard reaction (protein browning reaction) and result in the deterioration of freeze-dried product in storage. Consequently, sucrose and trehalose are the most widely used protective agents [81,82].

3. Polyols

The same as sugars, the functional groups of the polyols is hydroxyl as well. So polyol can protect biological products for freeze drying. Mannitol, sorbitol and glycerol are widely used polyols.

(1) Mannitol is white, odorless and sweet crystallization powder. It is easy to dissolve in water, but hardly dissolve in ethanol and ether. The melting point and boiling point of mannitol are 166—170℃ and 290—295℃, respecitvely. It is very stable in the sterile solution and difficult to be oxidized. Mannitol is usually used as bulking agent because it can form supporting structure for the active components by crystallizing during the slow freezing process. Mannitol does not react with the active components.

(2) Sorbitol is the isomer of mannitol, but its solubility is larger than mannitol. At room temperature, it is transparent viscous liquid that is optically active and hygroscopicity. It can dissolve many kinds of metal, but it is unstable at high temperature. Sorbital is normally used as bulking agent in freeze-drying.

(3) Glycerol (or glycerin) is viscous liquid that is transparent, odorless and sweet. Its melting point and boiling point are 17.9℃ and 290℃, respectively. The water absorption ability of glycerol is strong. Glycerol can mix with water and ethanol in any proportion. It is slightly soluble in ether and hardly soluble in organic solvent such as benzene, chloroform, carbon tetrachloride and carbon disulfide. But it is not soluble in fats. Glycerol is usually used as cryoprotective agents.

6.1.3 Effect of sugar/polyol concentration on the protection

The protective effect related with the concentration of sugar/polyol. Phosphofructokinase (PFK) is a kinase enzyme that catalyzes the irreversible transfer of phosphate from ATP to fructose-6-phosphate. Fig.6.1 shows the influence of the initial trehalose concentration on the protective results for freeze-dried PFK. It can be seen that the activity of rehydrated PFK is less than 40% when the concentration of trehalose is less than 100mol/L. The activity of PFK is more than 60% when the concentration of trehalose reaches 200—400mol/L.

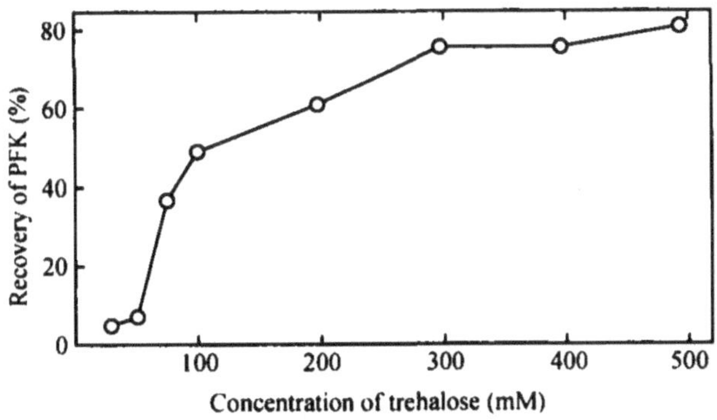

Fig.6.1 The influence of the initial trehalose concentration on the protective results for freeze-dried PFK [77]

Fig 6.1 shows that the recovery of PFK does not increase when the initial concentration of sugar-type protective agents reach a fixed value which is called the most effective concentration. Too high concentration of sugar/polyol even cause the protein degeneration. For example, in 1999, Allison reported that 5% sucrose can get the best protective effect for the freeze-drying of actin, while 10% (w/v) of sucrose have no obvious protective effect[83]. For freeze-dried protein, the recovery of protein is 65% if the concentration of trehalose is up to 150mg/mL. The recovery of activity will gradually decrease with the increment of trehalose concentration. Freeze dried protein will not have activity when the concentration of trehalose is 400mg/mL. But in the freezing-thawing experiment of protein, the recovery rate of the thawed protein will reach 90% when the concentration of trehalose is 400mg/mL, which means that high trehalose concentration lead to the activity loss in the drying process[84].

6.2 Polymer-type protective agents

6.2.1 The definition of polymer

Polymer is a large molecule (macromolecule) composed of repeating structural units typically connected by covalent chemical bonds. One polymer molecular consists of thousands to millions of atoms, so its molecular weight is large. Polymers are mainly classified as chain polymers and net polymers.

6.2.2 The polymer-type protective agents for freeze-dring

Polymers are used as protective agents in biological products' solution because they can increase the glass transition temperature of the solution. The common polymer-type protective agents are shown in Table 6.2.

Table 6.2 The common polymer-type protective agents[4]

name	molecular formula	relative molecular weight	glass transition temperature T_g' (℃)	collapse temperature T_{coll} (℃)
polyethylene glycol PEG	$H[OCH_2CH_2]_xOH$	$(2–400)\times10^2$		−13
dextran	$[C_6H_{10}O_5]_x$	$(1–200)\times10^4$		−10
hydroxyethyl starch HES			−12	>-5
ficoll		$(7–40)\times10^4$		-20
gum Arabic (acacia)		25×10^5		
gelatin				−8
polyvinylpyrrolidone PVP	$[CHN(CH_2)_4CO]_x$	$(3–36)\times10^4$		−24~−27
cellulose				
β-cyclodextrin	$C_{48}H_{80}O_{40}$	1135.00		
methocel		$(4–18)\times10^4$		−9
maltodextrin 860				
sephadex G200				−10
bovine serum albumin BSA		67000	−11	

Generally, the polymer-type protective agents have the following characteristics:

(1) Crystallize firstly during freezing process.

(2) Have higher surface activity.

(3) Produce steric hindrance effect among protein molecules.

(4) Increase viscosity of the solution.

(5) Increase the glass transition temperature significantly.

(6) Restrain the crystallization of excipients with small molecules (eg. sucrose).

(7) Keep the pH value of solutions.

In the freeze-drying process of biological products, polymer-type protective agents act as both cryoprotective agents and dehydration protective agents. The typical polymer-type protective agents are PVP, BSA, dextran, PEG and so on.

6.3 Surfactants and amino acid-type protective agents and other additives

6.3.1 Surfactant-type protective agents

1. Definition of surfactant

Surfactants are wetting agents composed of hydrophilic and oleophilic groups that can reduce the surface tension of a liquid and reduce the interfacial tension between two liquids. It is soluble in oil because of the C—H chains and soluble in water because of polar groups (—COOH, —OH). When these molecules are located at air-water or oil-water interface, the hydrophilic groups face the water phase, while the oleophilic groups point at gas and oil phase.

Surfactants can be divided into two main types: ionized and non-ionized. The ionized surfactants can be ionized into ions when dissolving in water.

In the freeze-drying of biological products, surfactants can reduce the denaturation during freezing and dehydration which is caused by the stress in ice-water boundary. The surfactants also can act as humid agents during rehydration process.

2. The typical surfactants in freeze-drying

In recent years, the surfactants have been using in low temperature storage and freeze drying of food, pharmaceutical and biological products.

Because the non-ionized surfactants have relatively lower critical micelle concentration (CMC), they are usually used with low concentration. The Tween group is the most popular non-ionized surfactants. The typical surfactants are shown in Table 6.3.

Table 6.3 The typical surfactants used in freeze-drying

Name
Tween 80
Triton X-100
sucrose fatty acid monoester
3-[(3-cholamidopropyl)-dimethylammonio]-1-propanesul fonate (CHAPS)
Hydroxypropyl-β-cyclodextrin (HP-β-CD)
sodium dodecanesulphonate (SDS)
Brij35, Brij30
Lubrol-px
Pluronic F127

6.3.2 Amino acid-type protective agents

1. The definition of amino acids

An amino acid is a molecule containing both amine and carboxyl functional groups. Amino acids are the building blocks of proteins. The α-Amino acid is the most important type. α-Amino acid consists of a amino, a carboxyl, a R group and a hydrogen atom that attach to the same α-carbon atom as shown in Fig.6.2.

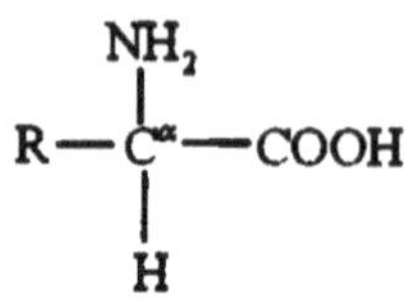

amino acid

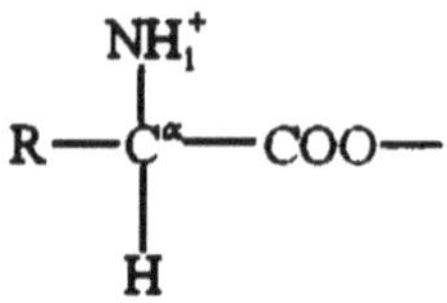

amine and carboxyl groups

Fig.6.2 Amino acid and its functional groups

Because amino acid ion is characterized by basic and acidic groups, it can inhibit the change of pH value of the solution during the low temperature storage and freeze-drying of biological products to protect the active components.

2. Amino acid-type protective agents

The typical amino acids used in freeze-drying are shown in Table 6.4. Because amino acids have both the amino and carboxyl groups, they can keep the pH value of solution. For example, glycine of low concentration can keep the pH value of sodium and potassium phosphate buffer of 10 mmol or 100 mmol to prevent the freezing denaturation of proteins[84]. Besides, some amino acids are useful bulking agents[85]. The glycine is the best bulking agent for the stability of freeze-dried IL-1 RA (Interleukin1, RA, pH6.5) stored at 8℃, 30℃ and 50℃ [86,91].

Table 6.4 The typical amino acid-type protective agents[4,85,86]

name	chemical formula	relative molecular weight	glass transition temperature T_g' (℃)
Proline	$(CH_2)_3NHCHCOOH$	115.13	
Glycine	CH_2NH_2COOH	75.07	-37
Glutamic acid	$(CH_2)_2NH_2CH(COOH)_2$	147.13	-17
Histidine	$NHCHNCHCCH_2CH(NH_2)COOH$	155.16	-32

continued

name	chemical formula	relative molecular weight	glass transition temperature T_g' (℃)
arginie	$HNC(NH_2)NH(CH_2)_3CH(NH_2)COOH$	174.20	
4hydroxyproline	$NHCH_2CH(OH)CH_2CHCOOH$	131.13	
L-serine	$HOCH_2CH(NH_2)COOH$	105.09	
β -alanine	$CH_2(NH_2)CH_2COOH$	89.09	-65
lysine hydrochloride	$H_2N(CH_2)_4CH(NH_2)COOH \cdot HCl$	182.65	
lysine	$H_2N(CH_2)_4CH(NH_2)COOH$	146.19	
sarcosine	CH_3NHCH_2COOH	89.09	
γ-aminobutyric acid	$H_2NCH_2CH_2CH_2COOH$	103.12	

6.3.3 Other additives

1. Antioxidants

During freeze-drying and storage process, the oxidizing reaction between O_2 and the position chain of Met, Cys, His, Trp and Tyr residues of protein can cause the deterioration of food and pharmaceutical products. In order to increase the stability and shelf life of freeze dried products, the appropriate amount antioxidants are added into the products to be freeze dried.

There are many types of antioxidants. The characteristics and mechanisms of antioxidation are different. Some types of antioxidants consume the O_2 inside the samples by self-oxidations to protect the sample from oxidizing. Some types of antioxidants inhibit the oxidation chain reaction of freeze-drying samples by providing electric and hydrogen atom. And some types of antioxidants restrain the activity of oxidase to prevent the oxidizing determination. The typical used antioxidants are shown in Table 6.5.

Studies show that the mixture of different antioxidants have better anti-oxidation effect than single antioxidant, even the mechanism is not clear[16].

Table 6.5 The typical used antioxidants[4,16]

name	chemical formula	relative molecular weight
antisterility factor (Vitamin E)	$C_{29}H_{50}O_2$	430.72
ascorbic acid (Vitamin C)	$C_6H_8O_6$	176.13
Lecithin	$C_{40}H_{82}NO_9P$	752.08
D(-)-Isoascorbic acid	$C_6H_8O_6$	176.13
L- ascorbic sodium	$C_6H_7NaO_6$	198.11
sodium thiosulfate anhydrous	$Na_2S_2O_3$	158.11
3-tert-Butyl-4 -hydroxyanisole	$C_{11}H_{16}O_2$	180.25

continued

name	chemical formula	relative molecular weight
butyl atea hydroxy toluene	$CH_3C_6H_2(OH)$ $[C(CH_3)_3]_2C_{15}H_{24}O$	220.36
propyl gallate	$(HO)_3C_6H_2COOCH_2CH_3$	212.20
ethylenediamine tetraacetic acid disodium salt dehydrate, Na_2EDTA	$C_{10}H_{14}N_2Na_2O_8 \cdot 2H_2O$	372.24

2. Buffer agents

Proteins can react with both acids and bases. The specific pH value, at which the number of positive charges is equal to the negative charges in the solution, is named the isoelectric point of the protein. The protein is more stable at its isoelectric point. At neutral pH values, most of the proteins carry negative charges while a few proteins carry positive charges. But at extreme pH values, the high electrostatical charge can cause strong intramolecular repulsion and split the proteins structures. Most of the protein denaturation induced by the pH value is reversible. However in some cases at basic pH, the peptide boding hydrolysis and accumulation can cause irreversible denaturation of protein.
The concentration of the protein solution increases gradually with the freezing of solution. High concentration of protein solution may change the pH values and cause the denaturation of the protein. So right amounts of buffer agents are usually added into the lyoprotectants. The typical buffer agents are shown in Table 6.6 .

Table 6.6 The typical buffer agents[52,86,94,121]

name	chemical formula	relative molecular weight	glass transition temperature T_g' (℃)	collapse temperature T_{coll} (℃)
citric acid monohydrate	$C_6H_8O_7 \cdot H_2O$	210.14	-54	
phosphoric acid	H_3PO_4	98.00		
ethylenediamine tetraacetic acid ,EDTA	$(CH_2)_2N_2(CH_2COOH)_4$	292.24		
tartaric acid	$C_4H_6O_6$	150.09		
4-(2-Hydroxyethyl)-1-piperazineethanesulfonic acid, HEPES	$C_8H_{18}N_2O_4S$	238.31	-63	
histidine	$NHCHNCHCCH_2CH(NH_2)$ COOH	155.16	-33	
potassium acetate	CH_3COOK	98.14	-76	

continued

name	chemical formula	relative molecular weight	glass transition temperature T_g' (℃)	collapse temperature T_{coll} (℃)
potassium citrate	$HOC(COOK)(CH_2COOK)_2$	306.42	–62	
potassium phosphate monobasic	KH_2PO_4	136.09		–55
sodium acetate	CH_3COONa	82.03	–64	
sodium bicarbonate	Na_2CO_3	105.99	–52	
sodium citrate	$HOC(COONa)(CH_2COONa)_2$	258.10	–41	
sodium phosphate	NaH_2PO_4	119.98	–45	
Tris base			–55	
Tris HCl			–65	

3. Agents for acceleration of freeze drying

Freeze drying is a time and energy consuming process. It is necessary to optimize the protocol of freeze-drying to shorten the freeze drying time. In recent years, it was found that needle-like crystal formed when tertiary butyl alcohol (TBC) was added into the lyoprotectants. The tube-shape channel formed with the sublimation of needle-like ice, resulting in the reduction of flowing resistance of vapor and increment of sublimating rate [87]. Such kind of additives as TBC is named as acceleration agents.

Tertiary butyl alcohol is a type of alcohol with small molecular. Its molecular formula is $(CH_3)_3COH$. It dissolves in water, has low toxicity and high vapor pressure. It can be used as freeze-drying solvent and compound solvent by combining with water. The compound solvent of tertiary butyl alcohol-water is being widely investigated nowadays [88,89].

Addition of tertiary butyl alcohol into the pharmaceutical solution has several purposes: ① reduce resistance of drying layer, accelerate freeze-drying process and shorten freeze drying time. ② dissolve pharmaceutical products that are difficult to dissolve in water. ③ increase the surface area of the products and make it easy to be rehydrated. ④ increase the stability of pharmaceutical solution and freeze-dried products. ⑤ inhibit the growing of bacteria.

For 5% w/v sucrose solution, it need 100 hrs (Fig.6.3a) to be freeze dried without tertiary butyl alcohol, while only 10 hrs (Fig.6.3b) are needed when 5% w/v tertiary butyl alcohol is added. The block-like ice crystals (Fig.6.4a) are not good for the escape of vapor. The needle-like ice crystals (Fig.6.4b) formed with the addition of tertiary butyl alcohol will form a channel when the ice was sublimitted. As a result, the flow resistance of vapor will be decreased greatly[88–90].

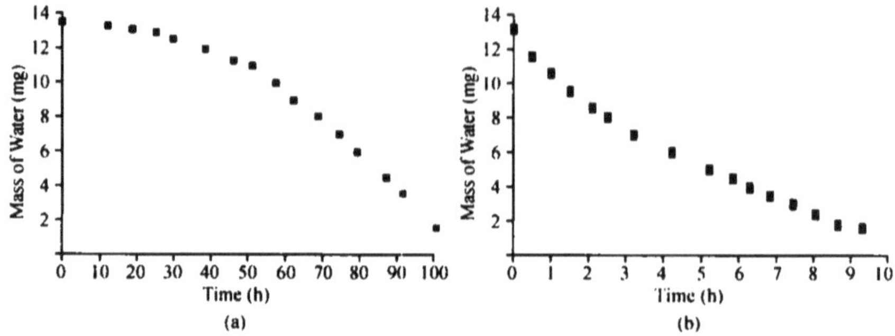

Fig.6.3 The change of water content with the freeze drying time [90]

(a) without TBC; (b) with TBC

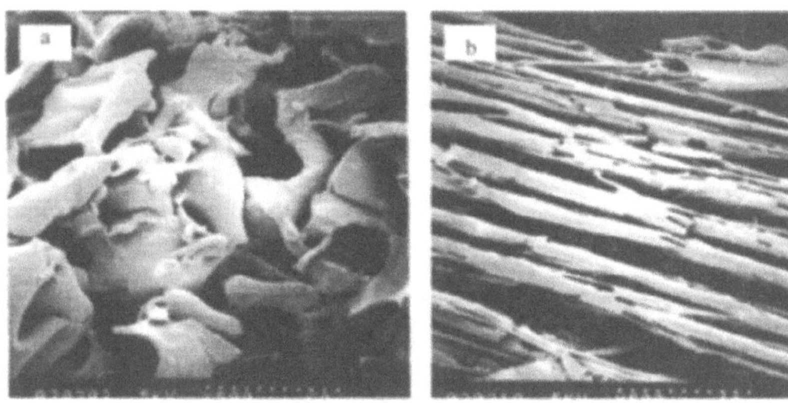

(a) (b)

Fig.6.4 The SEM photos of freeze dried sucrose solution

(a) without TBC; (b) with TBC

6.4 Formula issues in freeze-dried products

Because the active components of the biological materials are different, the formula is usually designed according to the products to be freeze dried. In order to obtain the highest recovery of activity, the compatibility among the various lyoprotectants, additives and the active components of biological products should be considered.

6.4.1 Determination of the optimal pH value

The active components of the biological products (such as protein) are stable only in very small range of pH value, and the pH values affect the solubility of some types of protein. For example, the proteins in IL-2 expanded and denatured irreversibly when IL-2 was freeze dried at pH7, the proteins kept their native structure when freeze dried at pH5.

In addition, the PH valve of the freeze-drying products will affect their long term storage and the physical/chemical stability of the solid-state protein. For example, the irreversible aggregation of the lyophilized RNase during the storage process was related to pH value of the protein solution before the freeze-drying process.

6.4.2 The selection of buffer agents

Many buffer agents can be used in the formula of biological products for freeze drying (as shown in Table 6.6). But each buffer agent can only be used to specific formula. For example, protein solution sensitive to pH should avoid using the buffer of sodium phosphate because Na_2HPO_4 crystallized first during the freezing process, inducing the decrease of pH and the denaturation of protein.

In addition, the correct selection of buffer concentration is also very important. For example, because of the crystallization of mannitol, it cannot protect the solution (2μg/mL) of β-galactosidase in the freeze-drying process. The addition of 10mM phosphate buffer (pH7.4) can partly inhibit the crystallization of mannitol, increasing the activity of enzyme to 95%. The crystallization of mannitol was completely inhibited when 200mM phosphate buffer (pH7.4) was added.

6.4.3 The selection of bulking agents

The addition of bulking agents (easy to crystallize) into the formula of freeze drying products will have additional functions as: ①providing adequate mechanical support for the final freeze-dried products. ②improving the appearance of the freeze-dried products. ③increasing the solubility of solutes. ④preventing the freeze-dried products from collapse or overflow.

The bulking agents should have the following characteristics: ①good solubility. ②compatetive with the active components of biological products.

③nontoxic. ④higher eutectic temperature. Mannitol and glycine are two typical bulking agents.

Some excipients may inhibit the crystallization of bulking agents, thus affecting the stability of biological products. For example, in the mixture of the sucrose and glycine, The amount of crystallization of glycine gradually decreases with the increment amount of sucrose. When the quality ratio of sucrose and glycine reached 1:1.4, the crystallization of glycine can not be observed. When the concentration of glycine is less than 29%, crystallization cannot be observed under the microscope, high quality freeze-dried products can be obtained only when the concentration increases to 50%. Similarly, in the mixture of mannitol with sucrose, trehalose, lactose or maltose, respectively, crystallization cannot be observed until the concentration of mannitol reached is greater than 30% (w/w).

6.4.4 The selection of cryoprotectants and lyoprotectants

In order to ensure long-term storage stability of freeze-dried biological products, Suitable cryoprotectants and lyoprotectants should be added.

Sugars are the widely used protective agent in the freeze-drying process of biological products because sugars won't crystallize at normal conditions of freeze drying. Reducing sugars are not chosen as protective agent since it may have react Maillard reaction with protein.

Sucrose is use to be the most widely used protective agent. In recent years, trehalose is being more applied because of its many advantages. It's difficult to compare their protective efficacy because it relates to many factors such as the types of biological products, concentration and so on.

Some salts can also be used as protective agent during the freeze-drying process of biological products. However, salts may decrease the glass transition temperature (T_g') of biological products. The factors influence the crystallization of salts include: the properties of the salts, the concentration of the salts and the cooling rate. According to their glass transition tendency at specific cooling rate and thermal history, Chang et al classified salts into three types: ①Crystal salt, such as maleic acid, Na_2HPO_4, Na_2SO_4, Na_2CO_3, KCl,$(NH_4)_2SO_4$. ②Partiy crystal salt (Unstable glass state), such as NaCl, $NaHCO_3$, K_2HPO_4, KH_2PO_4, $CaCl_2$, $MgCl_2$,Glycine, β-alanine. ③Glass transition salt, such as NaH_2PO_4, sodium/potassium citrate, citric acid, histidine, sodium/potassium acetate.

Some polymers is used as protective agents because they can increase glass transition temperature. However, the ability to form hydrogen bond between molecule of the polymers and protein is greatly lower than that between polymers and sugars. So, polymers and sugars are usually used together to not only increase the glass transition temperature of the formula but also form very strong hydrogen bond with the protein.

The active components of biological products (such as proteins)are also belonged to polymers, having high glass transition temperature. Therefore, improving the quality ratio between proteins and excipients can increase the glass transition temperature of freeze drying products and inhibit the crystallization of excipients as well. For example, the crystallization temperature of the sucrose solution with 4.4% water is about 70℃. Its crystallization temperature can be increased to 120℃ when containing 20% bovine growth hormone recombinant in sucrose solution[95]. The quality ratio between protein and sucrose is very high in the freeze-dried humicola lanuginose lipase because the proteins of high concentration inhibit the crystallization of sucrose. And the freeze-dried humicola lanuginose lipase is very stable at 60℃. The concentration of protein cannot be too high, otherwise protein will denature because there is not enough protective agents.

6.5 Examples of formulas

Generally, the contents of solid composition is between 2%—10% in the formula of protein. When the solid composition is less than 2%, hard freeze-dried products cannot be obtained. When the content of solid composition is more than 10%, it is difficult to completely dry the products, and even the freeze drying process is completed, the rehydration of freeze dried products is very difficult. The physical/chemical function between the additives is mutual. It is important to maintain the suitable amount of additives in the formula. If a type of additive can play several functions simultaneously, it will be the additives fo first selection. For example, some sugars can be used as cryoprotectants and lyoprotectants. In addition, large amount of buffer solutions or salts cannot be used because they may change the pH value during the freezing process or decrease the glass transition temperature of the formula. Some research results of the freeze-drying formula for protein and their protective efficacy in recent years were shown in Table 6.7.

Table 6.7 Examples of the formula for the freeze-drying of protein

protein name	formula	protective efficacy (measured)	experimental conditions
catalase, (ox liver)		activity retention	freezing at the temperature between -15～-70℃ for 10 minutes
	8.4μg/mL catalase, 10mM phosphate buffer(pH 7.0)	80%	
	+1M Glycerol	95%	
	+0.5mM Sodium Dodecyl Sulfate	90%	
elastase		activity retention	storage at the temperature 40℃ and the relative humidity is 79% for two weeks
	20mg/mL Elastase, 10mM sodium acetate buffer solution(pH5.0)	33%	
	+10%(w/v) Sucrose	86%	
	+10%(w/v) dextran 40	82%	
β-galactosidase		activity retention	freeze-drying
	20μg/mL β-galactosidase, 200mM phosphate buffer(pH7.4), 100mM Inositol	47%	
	+400mM Inositol	90%	
	+400mM Inositol, 1mg/mL dextran	100%	
	+400mM Inositol, 1mg/mL CMC-Na	100%	
recombinant factor XIII (rFXIII)		activity retention	freeze-drying
	2mg/mL recombinant factor XIII, 0.1mMEDTA、10mm Tris buffer(pH8)	82%	
	+100mM mannitol	90%	
	+1%(w/v)PEG	115%	
	+3.5%(w/v)dextran	109%	
	+100mM trehalose	117%	
	+100mM Sucrose	111%	
	+0.002%(w/v) Tween 20	106%	
H'-ATPase		activity retention	freeze-drying
	1mg/mL H'- ATPase、1mM EDTA-Tris buffer(pH7.0)	4%	
	+20mg trehalose /1mg protein	100%	
	+20mg maltose/1mg protein	91%	
	+20mg sucrose /1mg protein	84%	
	+20mg glucose /1mg protein	72%	
	+20mg galactose /1mg protein	37%	
rhIL-1ra		degenerate situation	storage at the temperature 50℃
	10mg/mL rhIL-1ra,2% aminoacetic acid, 10mM sodium citrate(pH6.5)	1.8%/week	
	+1% sucrose(w/v)	0.15%/ week	
	+1% maltose(w/v)	6.4%/ week	
	+1% sorbitol(w/v)	2.3%/ week	
	+1% trehalose(w/v)	0.43%/ week	

continued

protein name	formula	protective efficacy (measured)	experimental conditions
IL-2		activity retention	storage at the temperature 45℃ for 4 weeks
	0.8mg/mL freeze-driedIL-2、MES buffer(pH7)	47%	
	+0.5% sucrose(w/v)	86%	
	+5% mannitol(w/v)	28%	
LDH		activity retention	freeze-drying
	2μg/mL LDH, 50mM sodium phosphate buffer solution (pH7.4)	9%	
	+400mM mannitol	14%	
	+200mM Sucrose	30%	
	+10mg/mL CHAPS	76%	
	+10mg/mL HP-β-CD	73%	
	+10mg/mL Sodium Cholate,	60%	
	+10mg/mL PEG3000	30%	
	+10mg/mL PEG20000	38%	
PFK		activity retention	freeze-drying
	2mg/mL PFK, 20mM buffer solution of potassium phosphate(pH8.0)	0%	
	+1%PEG	20%	
	+10mM mannitol	5%	
	+10mM Lactose	5%	
	+10mM trehalose	0%	
	+1%PEG+10mM mannitol	30%	
	+1%PEG+10mM Lactose	103%	
	+1%PEG+10mM trehalose	100%	
hematopoietic stem cell		activity retention	freeze-drying
	+40% PVP(w/w)+20% sucrose (w/w)	31%	
	+10% Bovine serum	99%	
	+10% (w/w) mannitol	69%	
	+ 10% (W/W) mannitol + 10% Bovine Serum	98%	
HB-Ia liposome		particle Size before/after the freeze-drying process, μm	freeze-drying
	270mg HB-Ia liposome	8.21 / 5.89	
	+400mg sucrose	7.98 / 6.35	
	+800mg sucrose	6.58 / 5.33	
	+200mg sucrose +80mg glycine	8.58 / 5.83	
	+400mg sucrose +80 mg glycine	7.77 / 6.56	
	+ 400mg mannitol	7.89 / 6.24	
	+ 400mg PVP	6.44 / 7.08	
	+200 mg mannitol	8.21 / 5.89	

continued

protein name	formula	protective efficacy (measured)	experimental conditions
		entrapment efficiency after the freeze-drying	
ftorafur liposome	+10% glucose	25%	freeze-drying
	+10% sucrose	46.90%	
	+10% mannitol	45.40%	
	+10% trehalose	57%	

7

Freeze Drying of Pharmaceuticals

7.1 Basic issues for pharmaceutical freeze-drying

7.1.1 New Dosage forms of pharmaceutical drugs

Pharmaceutical preparations are used to prevent and cure diseases, and to guarantee the health and propagation of human beings. A safe, effective and stable dosage form for any pharmaceutical drug must be determined before clinical use. Pharmaceutics is the discipline of pharmacy that deals with all facets of the process of turning a new chemical entity(NCE) into a medication able to be safely and effectively used by patients. There are many chemicals with known pharmacological properties but a raw chemical is of no use to a patient. Pharmaceutics deals with the formulation of a pure drug substance into a dosage form.

According to its development, pharmaceutical dosage form can be divided into the following generations[9]:

The first generation are simple ointment, pill and powder for oral administration and external use.

The second generation are the tablets, injections, capsules and aerosols made by mechanical and automatic machines.

The third generation is slow-release or controlled-release dosage forms that form a new drug delivery system (DDS) .

The fourth generation is targeted dosage forms that form a targeted drug delivery system.

The fifth generation is the automatic release dosage forms inside the body when the patients have a serious illness.

At present, the third and fourth generations of dosage forms are most concerned by scientists.

In order to realize these new drug delivery system, many new techniques are developed in the formulation of dosage form, such as solid dispersion, inclusion, emulsion, liposomes, and microencapsulation.

Freeze-drying technology is widely used in the formulation of pharmaceutical dosage forms from the second generation to the fifth generation, especially in emulsion, liposomes and microencapsulation technology.

7.1.2 Biological drugs

1. Definition and classification of Biological drugs

Drugs can be divided into three categories: chemical drugs (or synthetic drugs); biological drugs; and natural drugs (Chinese medicine).

Biological drugs (or biological agents) are bioactivative preparations that are made from a living organism, parasites, animal toxins or their products and are used in the prevention, diagnosis, or treatment of cancer and other diseases. Biological drugs include antibodies, interleukins, and vaccines. Biological drugs can be classified according to the raw materials, biological or separation technology, or their clinical use.

According to biological technology, biological drugs can be divided into four categories [110]:

(1) Drugs produced by fermentation methods. Drugs are produced by microbial metabolism, including antibiotics, vitamins, organic acids, coenzymes, enzyme inhibitors, hormone, immunomodifier, as well as other physiologically active substances.

(2) Drugs produced by genetic methods. Protein and polypeptide-type drugs are produced by recombinant DNA, including interferon, insulin, interleukin-2 and so on.

(3) Drugs produced by cell engineering. Drugs are produced by cell culture of botanical and zoic cells, such as human physiological active factors, vaccine, monoclonal antibody, and so on;

(4) Drugs produced by enzyme engineering. They include medicinal enzyme and drugs which produced by enzyme or cells immobilization, such as protease, urgency enzyme, L-asparaginase, vitamin C, and so on.

The above methods are not entirely seperated. For instance, recombinant DNA technology can be used to improve microbial strains to produce better enzyme engineering drugs. Genetic engineering, cell engineering and traditional production techniques can be used together to select good strains and produce high efficient and low toxic broad-spectrum antibiotics.

Chemicals, biological drugs and Chinese medicine do not have clear boundaries. For example, Bio-pharmaceutical technology, such as fermentation and enzymatic technology, can be used to separate and extract effective components from Chinese herbal medicine.

Bio-pharmaceutical technology may provide valuable new drugs to the fetal diseases, such as autoimmune diseases (multiple sclerosis, arthritis, etc.), AIDS, cancer, senile psychosis, coronary heart disease, diabetes, and so on.

Biological drugs are in a stage of rapid development. According to the statistics of 2002, 133 biological drugs have been used in the clinic. Most of them was been approved within the 5 years since 1997 (FDA approved 19 drugs in 1997; 21 in 1998; 22 in 1999; 32 in 2000; 24 in 2001). There are about 350 biological drugs in the advanced stage of development [111]. According to the clinical use of drugs, biological drugs include vaccine, toxin, toxoid, serum, blood products, immune preparations, cytokine, antigen, monoclonal antibody, genetic engineering products (recombinant DNA products, external diagnostic reagents) and so on. At present, the major biological drugs are as follows: human serum albumin, erythropoietin(EPO), hemopoietin, interleukin, interferons, monoclonal antibody, vaccine, colony-stimulating factor, human growth hormone, insulin, cytokine, receptor drugs, blood coagulation factor Ⅷ, etc. New biological drugs being developed are tumor vaccine, polypeptide drugs, biotechnological external diagnostic reagents, etc.

2. Characteristics of biological drugs

Biological drugs have some distinct characteristics:

(1) The components are too complex to be measured accurately.

(2) Because the drugs are composed by a variety of protein that is active, they are affected by temperature. They are unstable and susceptible to denaturation and deactivation. The changes of parameters in production process have great influence on the quality of products[112] .

(3) they are susceptible to been polluted and destructed by microorganism.

Freeze-drying is a very important preparation method for the production of most biological drugs. According to the statistics of 1998, about 14% of antibiotic drugs, 92% of biomacromolecule drugs and 52% of other biological agents are freeze-dried. In recent years, the dosage forms of new biological drugs are all produced by freeze-drying technology. Moreover freeze-drying is the final technical procedure of drug production, which affect greatly the quality of drugs.

7.1.3 Basic process of biological drugs freeze-drying

The technical procedures of drug freeze-drying consist of four processes: preparation and freezing, primary drying (sublimation drying), secondary drying (desorption drying) and package. The temperature, vacuum for each process have to be controlled precisely. The freeze-dried drugs are dry and porous solids. They can be stored in room temperature or in refrigerator for a long time.

1. Preparation and freezing of drugs

In order to form a stable porous structure after freeze drying, the concentration of drug solution must be a specific value. Excipients should be added into the low dose thermal sensitive drugs (hormone, enzyme, vaccine) to reinforce the structure of freeze-dried products. Lyoprotectant should be added into the biological protein-type drugs or slow-release drugs with bio-membrane to protect proteins from denaturation and the bio-membrane from damage. The type and concentration of lyoprotectants are drug dependant. Most of the lyoprotectants also play the function of excipient. The quality of the freeze dried biological drugs is affected by the types, concentration and pH values of the lyoprotectants and excipients.

Freezing is a process to completely freeze or solidify the drug solution with controlled cooling rates. The end temperature of pre-freezing must be lower than the glass transition temperature (T_g) or eutectic temperature (T_e) of the drug solution.

2. Primary drying (sublimation drying)

Primary drying are performed at low temperature and vacuum. The drying progresses gradually from the surface to the center of the products. The pores or channels formed by the sublimation ice become the ways of vapor to escape. The boundary between drying layer and frozen layer is known as the sublimation interface. The temperature of the sublimation interface is a critical parameter to be controlled in primary drying process. 90% water in drugs is removed after primary drying.

In primary drying process, the temperature of frozen layer must be lower than Te or T_g'. The temperature of dried layer must is lower than the collapse temperature (T_c). The temperature of the heater in the drying chamber should be controlled strictly.

3. Secondary drying (desorption drying)

The purpose of secondary drying is to remove a portion of the bound water. The moisture content of drugs is lower than 3% after secondary drying. Because of large absorption energy, the product temperature in secondary drying must be increased high enough to remove the bound water, and on the other hand, this temperature cannot induce denaturation of proteins and deterioration of biological drugs. The T_g of the products increases gradually with the decrement of water in secondary process. So the drying temperature of the products can be increased gradually, but cannot be higher than T_g.

4. Encapsulation process

When the secondary drying process is complete, plugging system in the chamber is used directly to plug the vials in order to prevent the freeze dried drugs from oxidation and water absorption. The encapsulation can also be completed after filling nitrogen gas into the chamber.

7.1.4 Characteristics of freeze-drying technology for drugs

Compared with other drying methods, characteristics of freeze-drying technology for drugs are as fellow:

(1) Being carried at low temperatures, freeze drying can prevent the active components in drugs (bioprotein, hormone, vaccines, and so on) from denaturation or loss of biological activity.

(2) Being carried in a vacuum with extremely low oxygen content, freeze drying can protect the components in drugs from oxidation.

(3) Freeze-drying can greatly reduce the loss of volatile components in drugs.

(4) Freeze-drying can inhibit the growth of microorganism and the activity of enzyme in drugs.

(5) Freeze dried drugs will maintain the original structure.

(6) Freeze dried drugs have good rehydration property.

(7) Freeze dried drugs can be stored at room temperature for a long time because more than 95% of the water in drugs was removed

(8) The initial cost of freeze-drying equipment is larger. Freeze drying is a time and energy consuming process.

(9) It is very difficult to control the parameters at optimum level.

7.1.5 Critical problems of freeze-drying in drugs

There are three issues need to be considered in freeze-drying process of biological drugs. The first one is how to reduce the influence of freeze-drying on the efficacy of biological drug. The second one is how to control the optimum technical procedures of freeze-drying. The third one is how to reduce the time and energy consumption in freeze-drying process. Taking into account the above-mentioned issues, the critical technology involved in the freeze-drying of biological drugs can be summarized as the following:

1. Temperature Control and identification of drying procedures

It is very important to determine and control the optimum temperature in drying process. Frozen drugs will melt, collapse or crimple if the temperature is higher than the optimum temperature. While if the temperature is too low, refrigeration load will causing excessive energy consumption and the sublimation rate will be decreased greatly.

It is very important to identify the primary drying and secondary drying. If the secondary drying process starts earlier than the required time, frozen drugs will melt. More energy will be consumed if the primary drying last too long time. If the secondary drying process completes earlier than the required time, residual water of drugs will be too high and the shelf life of the dried drugs will be shortened. If the secondary drying lasts too long time, more energy will be consumed and the active components will deactivated due to excessive dehydration.

2. Cooling Rate in freezing process

Only ice forms when the temperature is greater than the eutectic temperature of aqueous solution. Solutes begin to precipitate when the temperature is lower than the eutectic temperature. The size of the ice crystals and the net structure of solutes are depending on the types of drugs and the cooling rates [113,114].

Consequently, freezing process determines the drying rate and the quality of freeze-dried product. The optimum cooling rates vary with different biological agents. For instance, slow freezing is usually beneficial to protein polypeptide-type drugs. Fast freezing is usually beneficial to the virus and vaccine. Therefore, the effect of cooling rates on the freeze drying of drugs should be investigated in the future.

3. Types and concentration of lyoprotectant

The molecular structure of the active components is different for different biological agents. The types and concentration of lyoprotectants required in freeze drying are also different. Up to now, there is not a universal lyoprotectant applied to all of the biological agents. It is an important task to develop the appropriate lyorotectants.

7.2 Freeze-drying of protein and hormone

Bio-protein-type drugs are typical products of modern bioengineering preparation. This type of drug is easy to be polymerized in liquid state because they are thermal sensitive. They are easy to be contaminated microorganism and oxidized when exposed in air. Freeze drying is an effective technique for the storage of bio-protein-type drugs. In this section, some basic issues in freeze drying of bio-protein-type drugs are illustrated by two examples.

7.2.1 Freeze-drying of enzyme

Almost all enzymes are proteins that catalyze chemical reactions. Ribonuclease is the polylyase used to catalyze ribonucleic acid (RNA). Townsend and De Luca[115] studied the function of lyoprotectants taking RNase as a protein model. The RNase solution is prepared in pH3~10 phosphate buffered solution. In the absence of lyoprotectant, Freeze-dried Rnase denatures in the form of covalent bond at 45 ℃. Ficoll 70 (water soluble ficoll) is an effective lyoprotectant for the freeze drying of RNase. The concentration increment of phosphate solution had a passive influence on the protein activity, because the protein molecule was polymerized by heavy metal ion. The increment of residual moisture content, increment of concentration of buffered solution and air trapped in the packaging vials all increase the loss of activity and denaturation of protein.

Chen investigated the protection of trehalose to the freeze-dried products containing enzyme[116]. He also investigated the protection of trehalose to serum alanine transaminase, aspartic acid aminopherase, γ-glutamyl transpeptidase, alkaline phosphatase, lactate dehydrogenase and creatinkinase in the freeze drying-rehydration process. The optimum concentration of trehalose was predicted.

Ge investigated the protection of trehalose to the Copper-Zinc Superoxide Dismutase (Cu/Zn SOD) [117]. The influence of different kinds of carbohydrates on the activity of Cu/Zn SOD was studied. The results showed that trehalose has the best protective property compared to other carbohydrates. The optimal amount of trehalose is 5%—10%(w/v). The activity of freeze dried Cu/Zn SOD can be preserved 100% with the addition of trehalose into the lyoprotectant. Ge also studied the effect of four typical buffer solution (Tris-HCl, PBS, sodium borate buffer, Na2ID04-citric acid) on the activity of Cu/Zn SOD. Results showed that the sodium borate buffer and Na2ID04-citric acid had better protect on the activity of SOD.

Nail[118] investigated the technical procedures of freeze-drying for lactate dehydrogenase. They found that supercooling degree had great influence on the activity of lactate dehydrogenase. The greater the supercooling degree, the smaller the activity. Sucrose can improve the activity of freeze dried lactate dehydrogenase. Glycine itself is not a good lyocryoprotectant for lactate dehydrogenase, but the mixture of sucrose and glycine in appropriate ratio can protect the activity of lactate dehydrogenase. The addition of glycine can increase the callapose temperature of the lactate dehydrogenase and inhibit the crystallization of sucrose in storage process. Surfactants such as Tween 80 or human serum proteins are conducive to improve activity of freeze dried lactate dehydrogenase. But the excessive addition of surfactant (especially Tween 80) is not conducive to the freeze-drying of lactate dehydrogenase.

7.2.2 Interleukin

Cytokine or growth factors are polypeptides used to regulate cellular proliferation and differentiation. Cytokine related to immunization is called interleukin. Interleukin-2 (IL-2)is a kind of cytokine excreted by lymphocyte to promote proliferation of T-lymphocytic.

Hora et al investigated the stability of human interleukin-2[119]. When used with amino acid, manicol and sucrose, interleukin-2 is very sensitive to pressure. When used with hydroxy-β- schardinger dextrin, interleukin-2 is more stable and more sentivitive to oxygen. When used with Tween 80, the products are more mechanically stable. If hydroxy-β- schardinger dextrin is contaminated or there is some water in Tween 80, interleukin-2 is more easily oxygenized.

Prestrelski et al.[120,121] investigated the influence of pH and lyoprotectants on the stability of interleukin-2 by infrared spectra. The results showed that pH value had great influence on the activity when there was no lyoprotectant added. When pH value is 7, the molecular structure of IL-2 is spreading. when pH value is less than 5, IL-2 can keep its natural structure which is more stable than that in pH 7. Lyoprotectant can change the stability of the drug products. The lyoprotectant can replace the position of water during drying process and protect the original molecular structure of IL-2. The lyoprotectant with higher glass-transition temperature can enhance the stability of freeze-dried drugs, but may not prevent the spreading of protein molecular structure.

Qi [122] investigated the lyoprotectants for human leukocyte interferon . It is shown that human serum albumin with concentration of 0.8%—1.0% is the best lyoprotectant. The biological activity of interferon was 100% recovered.

Interleukin -1, known also as lymphocyte activating factor, is a type of glycoprotein with molecular weight of 12 000—18 000. Byeong et al[123] investigated the glass temperature and storage stability of freeze-dried Interleukin-1 receptor antagonist with different lyoprotectant and buffer solution. Take sucrose, mannitol and glycine as lyoprotectant, they prepared several groups of samples. The results showed that if the sample was rehydrated immediately after freeze-drying, the receptor antagonist can be 100% recovered. If the freeze dried receptor antagonist (lyoprotectant is sodium phosphate solution containing 1% sucrose, 4% mannitol and 2% glycine, respectively) was stored at 40℃((glass transition temperature of the freeze-dried sample is 51℃), the activity loss of which is still relatively high. However, when the concentration of sucrose is increased to 5% in the lyoprotectant, the freeze dried acceptance antibody of recombinant human interferon can be preserved at the temperature below its glass transition temperature for a long time. Therefor ,glass transition temperature is the necessary condition, but not the sufficient condition to ensure the storage stability of freeze dried samples.

7.3 Freeze drying of fibrinogen

7.3.1 Fibrin glue

The properties of fibrin glue (also called fibrin sealant, FS) are better than any other biological or synthetic surgical hemostat in aspect of histocompatibility, nontoxicity and clinical efficacy. An electron microscope photograph of fibrin glue is presented in Fig.7.1.

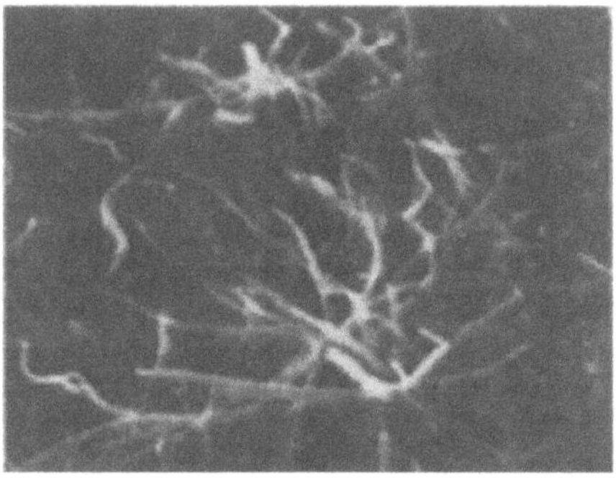

Fig.7.1 Electron microscopy photograph of fibrin glue (×6000)

Animal experiments confirmed that fibrin glue had five main pharmacological efficacy:

(1) stanching.

(2) bonding.

(3) closing up body cavity and promoting the repair of nerve.

(4) adherence function to increase the successful transplantation rate of skin graft.

(5) blocking exudate.

Substantial clinical trials indicated that fibrin glue can efficiently stop bleeding during the surgery process, particularly in controlling slow hemorrhage, fluid exudation, lymphatic exudation and substantive organs hemorrhage. It will be widely used in clinical surgery and endoscopic treatment.

The existing studies on fibrin glue mainly focuse on the clinical application of fibrin glue and the extraction, inactivation, gelation mechanism of fibrinogen. There are few reports about the freeze-drying of fibrinogen. In clinic the freeze-dried fibrinogen are required to rehydrate fast and to be stable in storage. At present, fibrin glue produced by the Immuno company (USA) is required to be stored at 2—8℃. Some fibrin glue produced in China can be stored at room temperature for less than a month. Some fibrin glue products need more than 30 minutes for rehydration, which could not meet the need of clinical application.

7.3.2 Fibrinogen

Fibrinogen is one of the two components of fibrin glue. It is a soluble plasma glycoprotein that is synthesised by the liver. It helps stop bleeding by helping blood clots to form. It is a blood coagulation factor which is extracted from blood plasma. Fibrinogen consists of two similar subunits, each subunits contains three polypeptide chains, $A\alpha$, $B\beta$ and γ . The N-terminal fragments of each subunit are linked by disulfide bridges; The N-terminal fragments of the six polypeptide chains curled into a intermediate, of which hydroxyls extend on both sides, forming a symmetric spheroids at the ends. Fibrinogen is rich in negative charges, making the molecules free. When combined with thrombin, fibrinogen can release fibrin peptide A and B from the N-terminal of $A\alpha$, $B\beta$ and eliminate central spheroid at the same time. The positive charges of the central spheroid and the negative charges of other spheroid at the ends are attracted by electrostatic. This enables the stepped polymerization among fibrin monomers. Then fibrin polymers form with the help of the hydroxyl and unstable hydrophobic bond. It is the fundamental structure of fibrous filaments in blood clotting. This is why fibrinogen is named. The molecular structure of fibrinogen is shown in Fig.7.2.

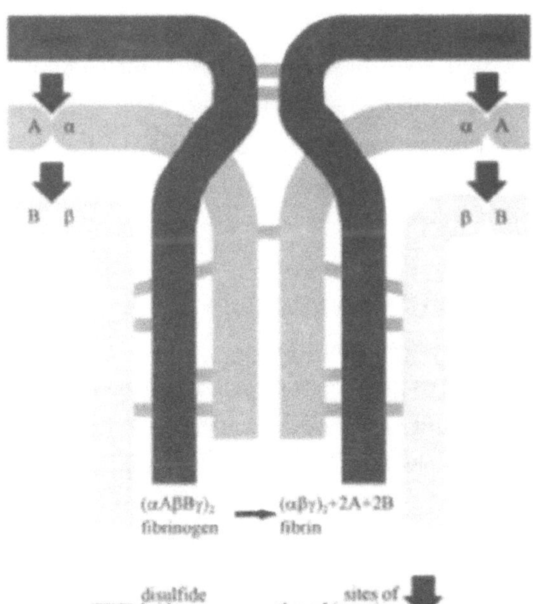

Fig.7.2 The molecular structure of fibrinogen

Fibrinogen molecular cracks fibrin peptide A and peptide B catalyzed by thrombin, forming fibrin monomer. Meanwhile, factor VIII activated by thrombin is involved in the fibrin monomer cross-linking. Fibrin monomer forms a stable and non-brittle clot.

Fibrinogen is a water-soluble active protein. It is unstable in liquid state, easily deteriorated by oxidation. When stored for a long time, it become yellow and condensative. Fibrinogen is thermal sensitive and will be denatured when heated. Fibrinogen needs to be freeze-dried and preserved by vacuum packaging. It is necessary to add appropriate lyoprotectant to increase the storage temperature and rehydration property of freeze-dried fibrinogen.

7.3.3 The freeze-drying of fibrinogen[124]

1. Fibrinogen preparation

1) Plasma Preparation

(1) Taking 5000mL of anti-coagulant (3.8% sodium citrate +0.9% sodium chloride + water), adding 5,000,000 of Gentamicin into it.

(2) Adding 45000mL of fresh pig blood into the above solution and mixing thoroughly;

(3) Standing the solution at 2—4℃ till blood cell precipitation and the interface appear;

(4) The supernatant is centrifuged at 3000r/min for 30 minutes at 2—4℃ .

(5) Supernatant of centrifuged solution is stored in a freezer at −30℃. Do not take out until it becomes plasma, with red-white color. This is the preparation for the extraction of fibrinogen.

2) Extraction of fibrinogen

(1) Taking out the frozen plasma from the freezer, and thawing at 2—4℃.

(2) Standing the thawed plasma at 2—4℃ for 2 hours, filtering and centrifuging at 1000rpm for 10 minutes.

(3) discarding the supernatant ,recording the volume of precipitation.

(4) Adding equal amount of fibrinogen washing solution (washing solution: sodium citrate 1.53g / L + NaCl 0.79g / L + Glycine 2.3g / L + glucose 3g / L, pH = 6.3).

(5) Centrifuging the mixture at 1000rpm for 10 minutes at 2—4℃.

(6) discarding the supernatant, calculating the volume of precipitation. The main component of precipitation is fibrinogen.

2. The temperature protocol for the freeze-drying of fibrinogen

Freeze-dryer (Lyo-7.5) is made in Shanghai Tofflon Science and Technology Co., Ltd. There are seven shelves and the area of each shelf 1200mm× 900mm.The shelf temperature can be adjusted in the range of −50—70℃. The temperature of condenser is −70℃. The ultimate vacuum pressure is less than 2.6Pa. Thermocouples are placed inside the vials on the tray as shown in Fig.7.3 during the freeze drying process.

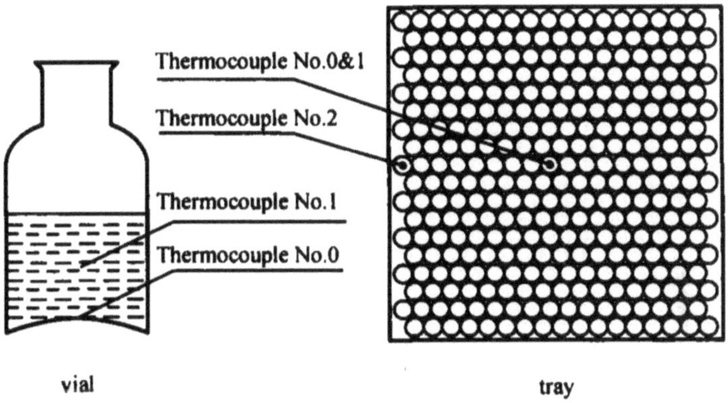

Fig.7.3 The location of thermocouples inside the vials on a tray

For the small vials located at the center of the tray, thermocouple No.0 is placed at the center of the bottom of the samples, while No. 1 is placed at the center of the vial. For the vials located at the periphery of the tray, No.2 thermocouple is placed at the center of the bottom of the samples. No.3 thermocouple placed on the shelf is used to measure the shelf temperature. The temperature - time curve of freeze drying for fibrinogen is shown in Fig. 7.4. The process of freeze drying is divided into three steps. Firstly, the shelf is cooled to –5℃ rapidly and hold at –5℃ for 20 minutes. Then the shelf is cooled to –50℃ quickly. When the temperature of the samples is cooled to –45℃ and held for one hour, the primary drying begins. The chamber is vacuumed to the pressure of 20 Pa and the shelf temperature is increased to –10℃. When the primary drying ends after 65 hours, the secondary drying begins. Finally, increase the shelf temperature to 5℃, holding for 5 hours and increase the shelf temperature to 35℃, holding for four hours. Then the whole freeze drying process ends. Pluging the vials under vacuum and then take out the vials for inspection.

It can be seen from Fig.7.4 that the samples' temperature was kept below −32 ℃ during the sublimation drying process. −32 ℃ is the collapse temperature of freeze-dried sucrose solution. Samples in the periphery of the tray (No. 2 thermocouple) has a higher temperature and higher drying rate than that at the center of the tray (No. 0 Thermocouple) because there is more radiation heat in the periphery of the tray.

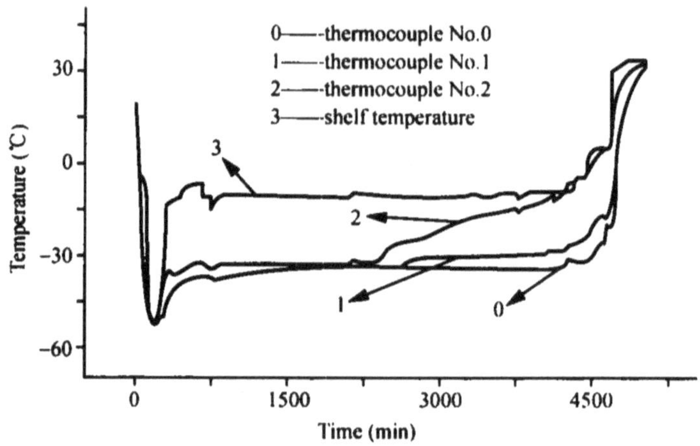

Fig.7.4 The temperature – time relationship during the freeze-drying process of fibrinogen

3. Lyoprotectant for the freeze-drying of fibrinogen

Sucrose, glycine and mannitol are combined to formulate the lyoprotectant. Sucrose, glycine and mannitol are common lyoprotectant for drugs. Sucrose is a type of antifreeze agent, that can protect the tissues and the molecular structure of protein from damage by crystals. Sucrose can enhance the stability of products during storage because its glass transition temperature after drying is relatively high. As bulking agents, glycine and mannitol have high eutectic temperature. They can keep the shape of the sample from shrinkage and increase drying rate. Sodium citrate solution and PBS solution are commonly used as buffer. It is reported that sodium citrate solution can enhance the storage stability compared with the PBS buffer solution because the former has higher glass transition temperature. A certain amount of sodium chloride and Tween 80 were added for the components of lyoprotetant. Sodium chloride can increase the ion concentration of solution. Tween 80 is a type of surfactant. They are used to prevent protein molecules from polymerization during freeze-drying.

The rehydration and stability of freeze dried products are different when the products are formulated with different concentrations of fibrinogen and lyoprotectant and different pH values of buffer (Sodium citrate or PBS with different concentration). To find out suitable lyoprotectants and buffers, 17 groups of samples were fromulated as shown in Table 7.1. 4000mL of liquid sample is put into four trays, with 250 vials (volume 7mL) on each tray. 4mL samples were filled in each vial.

Table 7.1 The preparation of fibrinogen solution for freeze-drying and its rehydration property

groups	fibrinogen (mg/mL)	lyoprotectants (mg/mL)			buffer	sodium chloride (mg/mL)	tween 80 (mg/mL)	rehydration time(min)
		sucrose	glycine	mannitol				
group 1	50			20	PBS(0.2mol, pH=7.9)	1	0.2	2
group 2	60		15		sodium citrate 4 mg/mL, pH=8.2	2	0.2	3
group 3	50	10	20		sodium citrate 4 mg/mL, pH=8.2	2	0.2	3
group 4	50	10		20	sodium citrate 4 mg/mL, pH=8.2	2	0.2	2
group 5	50	10		40	sodium citrate 6 mg/mL	2		2
group 6	40	30			PBS(0.2mol, pH=7.9)			6
group 7	50	30			PBS(0.2mol,pH=7.9)			8
group 8	50	50			PBS(0.2mol,pH=7.9)		0.2	8
group 9	50	30	5		sodium citrate 4 mg/mL, pH=8.2	2	0.2	2
group 10	50	30	15		sodium citrate 4 mg/mL, pH=8.2	2	0.2	3
group 11	50	30	20		sodium citrate 4 mg/mL, pH=8.2	2	0.2	3
group 12	50	30	10		sodium citrate 8mg/mL	2	0.2	6
group 13	50	30		20	sodium citrate 4 mg/mL, pH=8.2	2	0.2	2
group 14	72.6	10	20		sodium citrate 6mg/mL	1.34	0.3	20
group 15	77.2	50	20		sodium citrate 6 mg/mL	1.34	0.3	20
group 16	73.8	10	20	40	sodium citrate 6 mg/mL	1.34	0.3	20
group 17	70.4	50	20	40	sodium citrate 6 mg/mL	1.34	0.3	20

Notes: for Group 14 to 17, diluted hydrochloric acid is added to the formulars to adjust pH to 6.9.

4. The rehydration property and stability of freeze dried fibrinogen

At room temperature (25 ℃), taking 2mL of 0.2 mol/L, pH 6.9 PBS buffer using syringe and injecting it into the vial containing freeze-dried fibrinogen. Then rotating the needle to eliminate vacuum. Shaking the vial slightly and let it standing. Calculating the total time needed to completely dissolve (i.e, rehydration time). The results are also shown in Table 7.1.

7.4 Freeze-drying of liposome

7.4.1 Liposome and liposome drug

1. Liposome

A liposome is a tiny bubble (vesicle), made out of the same material as a cell membrane. Membranes are usually made of phospholipids, which are molecules that have a head group and a tail group. The head is attracted to water, and the tail, which is made of a long hydrocarbon chain, is repelled by water. A liposome has the function of biomembrane, like amphiphilic and mobility[9,125,126]. Liposomes were first described by British hematologist Dr Alec D Bangham in 1961 (published 1964), at the Babraham institute, Cambridge. Liposome is discovered when Bangham and R. W. Horne were testing the institute's new electron microscope by adding negative stain to dry phospholipids. The resemblance to the plasmalemma was obvious, and the microscope pictures served as the first real evidence for the cell membrane being a bilayer lipid structure. Nowadays, liposomes mostly mean artificial lipid vesicles.

The correct choice of liposome preparation method depends on the following parameters: ① the physicochemical characteristics of the material to be entrapped and those of the liposomal ingredients; ② the nature of the medium in which the lipid vesicles are dispersed; ③ the effective concentration of the entrapped substance and its potential toxicity; ④ additional processes involved during application/delivery of the vesicles; ⑤ optimum size, polydispersity and shelf-life of the vesicles for the intended application; ⑥ batch-to-batch reproducibility and possibility of large-scale production of safe and efficient liposomal products. It should be noted that formation of liposomes and nanoliposomes is not a spontaneous process. Lipid vesicles are formed when phospholipids such as lecithin are placed in water and consequently form one bilayer or a series of bilayers, each separated by water molecules, once enough energy is supplied. Liposomes can be created by sonicating phospholipids in water. Low shear rates create multilamellar liposomes, which have many layers like an onion. Continued high-shear sonication tends to form smaller unilamellar liposomes (Fig. 7.5). In this technique, the liposome contents are the same as the contents of the aqueous phase. Sonication is generally considered a "gross" method of preparation as it can damage the structure of the drug to be encapsulated. Newer methods such as extrusion and Mozafari method are employed to produce materials for human use.

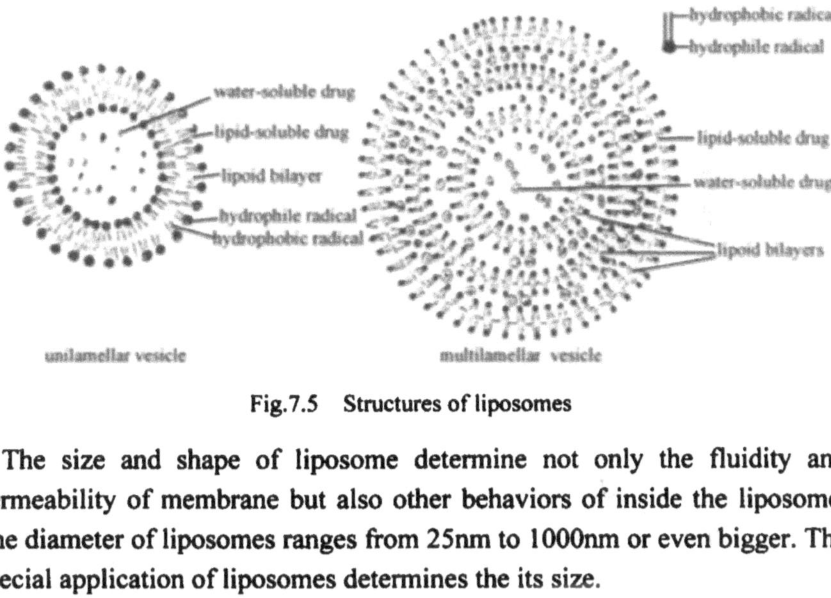

Fig.7.5 Structures of liposomes

The size and shape of liposome determine not only the fluidity and permeability of membrane but also other behaviors of inside the liposome. The diameter of liposomes ranges from 25nm to 1000nm or even bigger. The special application of liposomes determines the its size.

Liposomes are used for drug delivery due to their unique properties. A liposome encapsulates a region on aqueous solution inside a hydrophobic membrane; dissolved hydrophilic solutes cannot readily pass through the lipids. Hydrophobic chemicals can be dissolved into the membrane, and in this way liposome can carry both hydrophobic molecules and hydrophilic molecules. To deliver the molecules to sites of action, the lipid bilayer can fuse with other bilayers such as the cell membrane, thus delivering the liposome contents. By making liposomes in a solution of DNA or drugs (which would normally be unable to diffuse through the membrane) they can be (indiscriminately) delivered past the lipid bilayer. There are three types of liposomes- MLV (multilamillar vesicles), SUV(Small Unilamillar Vasicles) and LUV(Large Unilamillar Vesicles). These are used to deliver different types of drugs.

Liposomes are used as models for artificial cells. Liposomes can also be designed to deliver drugs in other ways. Liposomes containing low (or high) pH can be constructed such that dissolved aqueous drugs will be charged in solution (i.e., the pH is outside the drug's range). As the pH naturally neutralizes within the liposome (protons can pass through some membranes), the drug will also be neutralized, allowing it to freely pass through a membrane. These liposomes work to deliver drug by diffusion rather than by direct cell fusion. Another strategy for liposome drug delivery is to target endocytosis events. Liposomes can be made in a particular size range that makes them viable targets for natural macrophage phagocytosis. These liposomes may be digested while in the macrophage's phagosome, thus releasing its drug. Liposomes can also be decorated with opsonins and ligands to activate endocytosis in other cell types.

In addition to gene and drug delivery applications, liposomes can be used as carriers for the delivery of dyes to textiles, pesticides to plants, enzymes and nutritional supplements to foods, and cosmetics to the skin.

The use of liposomes in nano cosmetology also has many benefits, including improved penetration and diffusion of active ingredients, selective transport of active ingredients, longer release time, greater stability of active, reduction of unwanted side effects, and high biocompatibility

2. Characteristics of liposomal drug [127—129]

As for its unique structure, liposome has following features when used as drug carriers:

(1) Targeting. One of the most prominent properties of liposomes is the targeting. The targeting means the directional distribution of drugs in human body. The phagocytotic endothelial cells can engulf most of liposomes inside the human body. If the tumor cells have phagocytotic function, liposomes will target at tumor cells as well. For this reason, liposomes are used in the treatment and prevention of proliferation and metastasis of tumor cells.

(2) Slow-releasing. The retention of liposomal drugs in blood circulation is much longer than that of free drugs. The retention time of different liposomal drugs can vary between a few minutes to several days. Liposomes with different half-life can be designed as drug carriers to release the drugs according to the requirement of patients, improving therapeutic efficacy.

(3) Reducing the toxicity of drug. After encapsulated by liposome, drugs are mainly engulfed by phagocyte in reticuloendothelial system of the liver, spleen and bone marrow. Drugs released to heart and kidney is much less than that of free drugs. Therefore, if some drugs are particularly toxic to healthy cells of heart and kidney, drugs encapsulated in liposomes can significantly reduce their toxicity .

(4) enhancing the stability of encapsulated drugs. It's validated by experiments that once some unstable drugs are encapsulated in liposomes, their stability will be significantly improved due to the protection of liposome bilayer.

3. Some problems in the application of liposome drug

Although liposome can encapsulate both water-soluble drugs and fat-soluble drugs and have good histocompatibility due to the similar composition with human cell membranes, it has some shortcomings which limits its clinical application.

Liposome is unstable and vulnerable to the changing pH value and the drug encapsulated. It can only be stored for several weeks in liquid state. Some drugs are prone to be hydrolyzed or damaged by high-temperature. When stored with liposomes in liquid state, these drugs will lose their efficacy. The fat in liposome is the ideal culture for many bacteria, so liquid liposome is easy to be contaminated by bacteria.

The structure of liposome is greatly affected by temperature. Liposomal structure will become unstable when temperature rises above 50 ℃ and phospholipid bimolecular chain structure will be undermined. Liposome is mainly constituted by phospholipid, which is very easy to be oxidized. Oxidation not only affects the drug entrapment efficiency but also produces lysophosphatidic acid which is harmful to human body. Liquid liposome is a suspended emulsion, which can easily agglomerate, amalgamate and lead to drug leakage.

In order to realize the long term storage of liposomes and the encapsulated drugs at room temperature, freeze-drying technology is being widely used in the preparation of liposome drug [96,130–132].

7.4.2 Preparation method of liposomes

As an example, a liposome with 5% (w/v) glucose used for lyoprotectant is intruduced. Balance 1.2g soybean lecithin, 0.6g cholesterol, 0.4g poloxamer and put them into a sprayer. Then pour 10mL chloroform ($CHCl_3$) into the sprayer and mix them to ensure dissolving the materials in the sprayer. Put 2.0g glucose into coating pan, then rotate the coating pan. Spray the membrane materials in the sprayer into the coating pan, twice every 10 minutes. Blow the coating pan with a warm air blower to volatile the organic solvent inside membrane material. After all membrane materials are sprayed out, coating pan continues to rotate for 20 minutes. Take the formulated liposome out of coating pan and put into small bottles, seal the bottles and store them in a desiccator. Add distilled water into the small bottle to get a 40mL suspension, which is liposome with 5% (w / v) glucose. The same method can be used to formulate other suspension of liposome with glucose, sucrose, mannitol, trehalose.

7.4.3 The measurement of T_g of liposome suspension [133]

The glass transition temperature (T_g) of liposome suspension with different protectants were measured by DSC (DSC-204, NETZSCH), as shown in Table 7.2.

Table 7.2 The glass transition temperature (T_g) of liposome suspensions (The samples were cooled to −65℃, holding for 10 min at −65℃, and then heated at a rate of 10℃ / min)

experimental conditions	T_g (℃) glucose as protectant	T_g(℃) sucrose as protectant	T_g (℃) mannitol as protectant	T_g (℃) trehalose as protectant
concentration of protectant: 15% cool rate: 5 ℃ / min	−38.5	−32.4	−30.4	−29.2
concentration of protectant: 15% cool rate: 5 ℃ / min	−39.8	−33.8	−31.7	−31.4
concentration of protectant: 15% cool rate: 10 ℃ / min	−38.7	−32.9	−30.9	−30.7

7.4.4 The freeze-drying process of Liposome [134–136]

In order to study the effects of the technical parameters on the quality of drugs in the freeze-drying process, the temperature of heating shelf, the temperature of cold trap and the vacuum degree must be measured. A freeze-drying experimental system with a electronic balance inside the drying chamber was designed and set up.

The liposomal suspensions were frozen to −65℃ at different cooling rates and put into the drying chamber of freeze dryer quickly. The vacuum degree of the drying chamber is 10Pa.

Freeze-drying temperature - time curves are shown in Fig.7.6. Mass - time curves are shown in Fig.7.7.

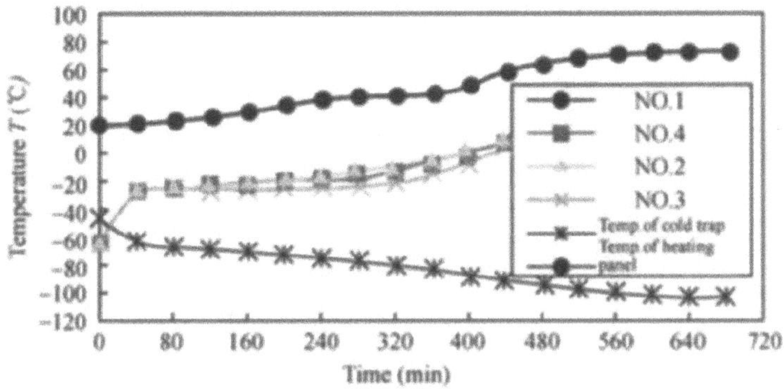

Fig.7.6 Freeze-drying temperature - time curves of a tsting materials of liposome (Point one, two, three means the thermalcouples)

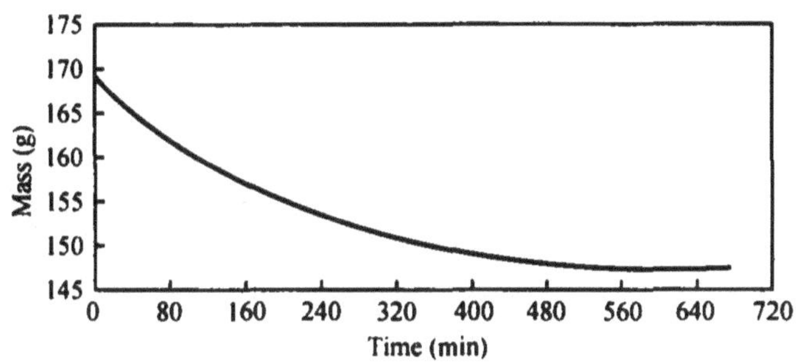

Fig.7.7 Mass loss of a testing material of liposome during drying

7.4.5 The change of the size of liposomel particles before-and-after drying

The size of liposomal particles has significant effect on their functions in vivo, such as the position at which act with cells and the distribution, the absorption in human body. For example, after intravenous injection, small-size liposomes can reach hepatocytes quickly, medium-size liposomes can maintain a certain time in the circulation of blood, while large-size liposomes can not reach the drug targets. Andrew et al. reported that vesicle size has become an important index to judge the protective effect of liposomes.

The sizes of liposomal particles were measured by using of the TSM-type Ultrafine Particle Analyzer. The distribution of liposomal particle size with 15% trehalose as protectant is shown in Fig.7.8. The average particle size of liposomes with trehalose of different concentrations is shown in Fig.7.9.

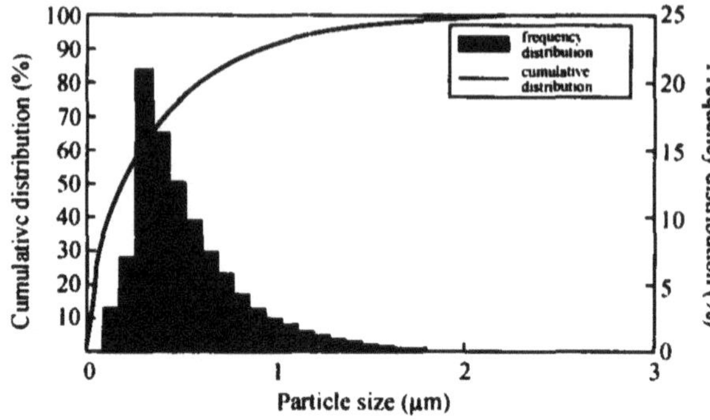

Fig.7.8 The distribution of liposomal particle size with 15% trehalose as protectant

The change of particle size of liposome is taken as a typical parameter to evaluate the freeze drying qulity of liposomes. If the particles of liposome blend with each other during the freezing and drying process, the particle size and distribution of liposome change distinctly. It can be seen from the results the change of particle diameter of liposome is smallest when trehalose is the protectant. While the particle diameter changes is the largest when glucose is the protectant. 10% trehalose is the most effective lyoprotectant for liposome. For mannitol, sucrose and glucose, the optimum concentration for liposome freeze drying is 15%, 10% and 15% , respectively.

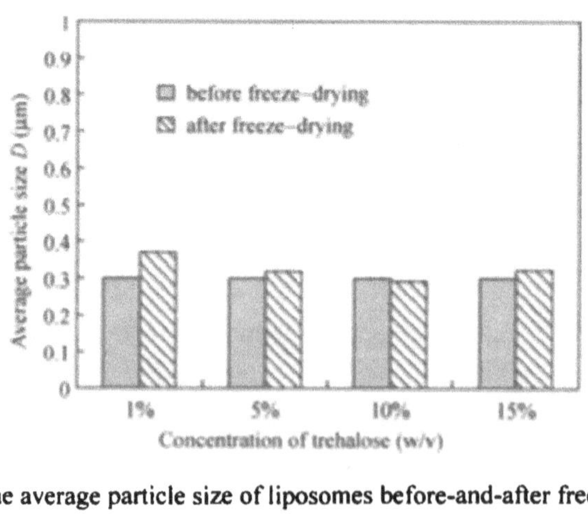

Fig.7.9 The average particle size of liposomes before-and-after freeze-drying (trehalose as protectant)

7.4.6 The effect of cooling rate on the particle size of freeze dried liposome

In order to study the effects of cooling rate on the particle size of liposome, liposomes with 10% trehalose were cooled to –65℃ by using programmer controlled freezer at the rates of 1℃/min and 20℃/min respectively. The frozen liposomes were then dried in a freeze dryer.

In the drying process, vacuum degree was kept at 10Pa. The temperature of heating shelf was controlled during the primary drying. Because the free water in liposomes is easy to be removed, it is enough to keep the temperature of cold trap at –60 ℃. During the secondary drying, the bound water is comparatively hard to eliminate. In order to shorten the drying time, the temperature of the heating shelf should be increased and the temperature of cold trap should be reduced to –100 ℃ to increase the vapor pressure difference between liposome and the cold trap.

The temperature and moisture change of the liposome which was frozen by different cooling rates are showed in Fig.7.10 and Fig.7.11, respectively.

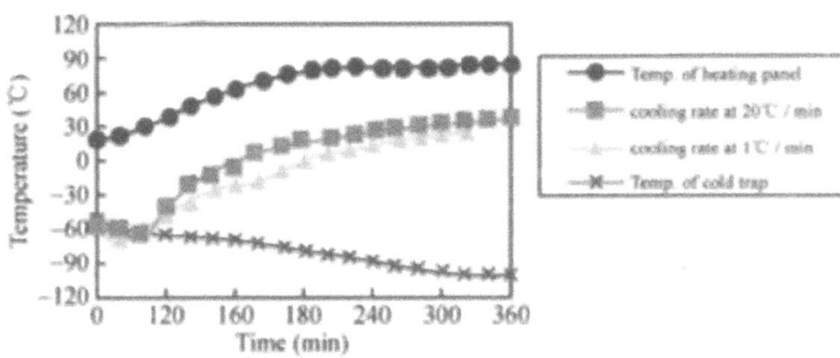

Fig.7.10 The temperature change of the liposome frozen by different cooling rates

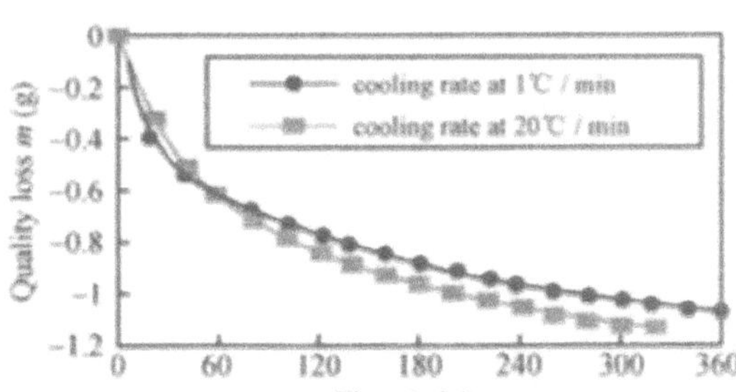

Fig.7.11 The moisture change of the liposome frozen by different cooling rates

Before freeze-drying, the average particle diameter of liposomes is 281nm. The average particle diameter of the liposome rehydrated is 456nm when cooled at the rate of 1 ℃/min, and 298nm when cooled at the rate of 20℃/min. The changes of particle diameter of liposome before-and-after freeze-drying at different cooling rate are shown in Fig.7.12.

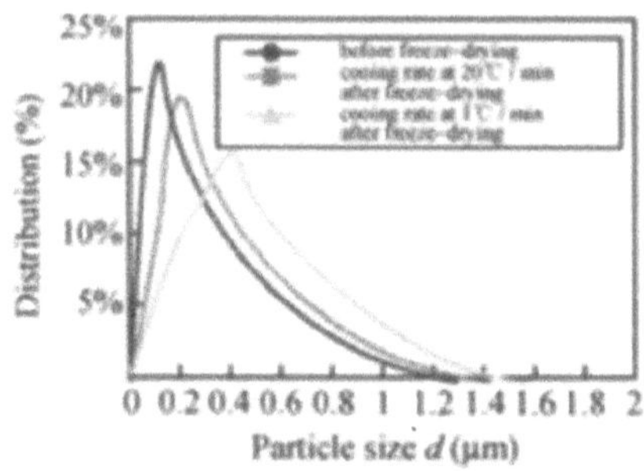

Fig.7.12 Diameter distribution of liposomes frozen at different cooling rates

7.5 Freeze-drying of water-soluble and fat-soluble liposome pharmaceuticals

Both water-soluble and fat-soluble pharmaceuticals can be encapsulated by liposomes. In order to study the freeze-drying of liposome pharmaceuticals, ftorafur (water-soluble) and vitamin A(fat-soluble) were selected to study the freeze drying of liposome pharmaceuticals[130,134,137].

7.5.1 Preparation of ftorafur liposome

Ftorafur is the derivative of 5-fluorouracil. Catalyzed by liver microsomal enzyme, ftorafur is breaks down into 5-FU, a substance that kills tumor cells. Ftorafur is a potential drugs for the treatment of some types of cancers. In order to enhance the drug efficacy and reduce toxicity to healthy tissues, ftorafur is formulated into the form of liposome to maintain a higher concentration in targeted tissue.

1. The preparation of ftorafur liposome by rotating volatile method

Preparation of ftorafur liposomes with 5% (w/v) glucose as lyoprotectant is intruduced. Balance 1.2g soybean lecithin, 0.8g ftorafu, sieveing twice (100 mesh) and mixing thoroughly. Balance 1.2g soybean lecithin, 0.6g cholesterol, 0.4g poloxamer and put them into a sprayer. Then pour 10ml chloroform

($CHCl_3$) into the sprayer and mix to ensure dissolving the materials in the sprayer. Put the mixtures of glucose and ftorafur into coating pan, then rotate the coating pan. Spray the membrane materials in the sprayer into the coating pan, twice every 10 minutes. Blow the coating pan with a warm air blower to volatile the organic solvent inside membrane material. After all membrane materials are sprayed out, coating pan continues to rotate for 20 minutes. Take the formulated liposomes out of coating pan and put into small bottles, seal the bottles and store them in a desiccator. Add distilled water into the small bottle to get a 40mL suspension, which is ftorafur liposomes suspensions with 5% (w / v) glucose. The same method can be used to formulate other suspension of ftorafur liposomes with glucose, sucrose, mannitol, trehalose.

2. The preparation of ftorafur liposome by Ultrasonic Irradiance method

Liposome suspensions with different lyoprotectants (glucose, sucrose, mannitol, trehalose) can also be formulated by other methods, such as ultrosonic irradiance method.

Preparation of ftorafur liposomes with 10% (w/v) glucose as lyoprotectants is intruduced as below. Balance 0.6g glucose, 0.12g ftorafur, 0.12g phosphoric acid, 0.06g cholesterol and 0.04g poloxamer and put them into a volumetric flask. Then add 6ml buffer solution containing sodium dihydrogen phosphate and disodium hydrogen phosphate and 0.2ml ethanol.

In order to fully dissolve the mixtures, the volumetric flask is immersed in a constant temperature bath at 60℃ for 5min and then put on an ultrasonic emulsifier (BRANSON SONIFIER-450). Under high pressure, the liposome suspension sprays on the edge of resonator, arising vibration at resonant frequency of the edge and the vibration up and down of fluid flow. If the ultrasonic frequency is high enough, the fluid flow will be emulsified into fine emulsion as a result of high frequency vibration. The ftorafur liposome suspension with the glucose concentration of 10% (w / v) will be obtained under the ultrasonic frequency of 30 kHz for 8min. The same method can be used to formulate other suspension of ftorafur liposomes with glucose, sucrose, mannitol and trehalose.

7.5.2 Preparation of vitamin A liposome

A vitamin is an organic compound required as a nutrient in tiny amounts by an organism. Vitamin A, a general designation of all ionone derivatives with retinol biological activity, is soluble in fat and fat solvent but not in water. A lack of vitamin A will lead to night blindness, keratomalacia, epidermal cells keratinization and so on. It is sensitive to air, oxidant and UV. Its decomposition rate is accelerated with increasing temperature. The cosmetics made of vitamin A liposome can reduce the oxidative deterioration of vitamin A and then have more efficacies to human health.

Vitamin A is light yellow viscous liquid. Vitamin A liposome suspension can be formulated by ultrasonic irradiance method.

Put 0.6g lyoprotectant, 30μL vitamin A, 0.12g phosphoric acid, 0.06g cholesterol and 0.04g poloxamer into a volumetric flask, then add 6mL buffer solution with sodium dihydrogen phosphate and disodium hydrogen phosphate and 0.2mL ethanol. In order to fully dissolve the mixtures, the volumetric flask is immersed in a constant temperature bath at 60℃ for 5min and then put on an ultrasonic emulsifier for 8min. Therefore, the vitamin A liposome suspension with the glucose concentration of 10% (w / v) is obtained.

7.5.3 Freeze-drying of liposome pharmaceuticals

The liposome pharmaceuticals are cooled to –65℃ at 20℃/min by a programmable freezer. Then the frozen liposome pharmaceuticals are transfer to the freeze dryer quickly. The temperature of the pharmaceuticals was controlled at below 30℃ by regulating the temperature of heating shelf. The technical procedure of liposome pharmaceuticals is similar with that of liposome as introduced in the last section.

7.5.4 Measurement of encapsulation of liposome pharmaceuticals

Encapsulation is an important parameter to judge the quality of freeze dried liposome pharmaceuticals. Encapsulation can be expressed as

$$E_n = \frac{C_t - C_f}{C_t}$$

where C_t——quantity of pharmaceuticals in suspensions of liposome;

C_f——quantity of free pharmaceuticals in suspensions of liposome.

1. Measurement of ftorafur liposome encapsulation

Instruments: Waters HPLC (C18 column. column temperature: 24℃; column diameter: 4.6mm; column length: 150mm; mobile phase: 30% methanol-water solution; flow rate: 1.0mL/min; detection wavelength: 270nm;

sensitivity: 0.05AUFS; paper speed: 10mm/min); LD5-2 electric centrifuge (Beijing Medical Centrifuge Factory),with the maximum speed of 5000RPM.

1) Standard curve

Prepare methanol solution of 10mL with the ftorafur concentration of 1μg/mL, 2μg/mL, 3μg/mL, 4μg/mL, 5μg/mL, 6μg/mL respectively. Pipette 10μL prepared solution into the chromatograph through the injection valve. Each solution was measured 7 times in order to obtain the ftorafur chromatogram. The peak area was calculated by use of the chromatogram. Draw a concentration of ftorafur-peak area curve and perform a linear regression analysis to obtain the standard curve for the calculation of ftorafur concentration.

2) Measurement of the content of free ftorafur C_f

Add 500μL liposome suspension with ftorafur into a centrifugal tube and centrifuge at a speed of 3000r/min. Take 6.5μL supernants into a 25mL volumetric flask and then add methanol till the liquid level reach the 25mL mark line. Pipette 10μL of the prepared solution into the chromatograph through a injection valve. Each solution is measured twice. The content of free ftorafur can be calculated by the standard curve of ftorafur concentration.

3) Measurement of the total amount of ftorafur

Take 6.5 μL liposome suspensions into a 25mL volumetric flask and then add methanol till the liquid level reach the 25mL mark line to damage the liposomes. Pipette 10μL of the prepared solution into the chromatograph through a injection valve. Each solution is measured twice.

The encapsulations of ftorafur liposome measured before and after freeze-drying is showed in Table 7.3 to Table 7.5.

Table 7.3 Change of encapsulation of ftorafur liposome prepared by rotary volatile method (rapid cooling: cool down to –65℃ with a cooling rate of 20℃/min)

lyoprotectant	encapsulation before freeze-drying (%)	encapsulation after freeze-drying (%)
5% glucose	44.2	21.3
10% glucose	43.5	25.0
15% glucose	44.6	36.8
5% sucrose	53.2	37.6
10% sucrose	51.3	46.9
15% sucrose	49.4	48.7
5% mannitol	51.4	39.2
10% mannitol	49.3	45.4
15% mannitol	48.7	46.8
5% trehalose	58.9	53.6
10% trehalose	59.2	57.0
15% trehalose	61.7	60.5

Table 7.4 Change of encapsulation of ftorafur liposome prepared by rotary volatile method

lyoprotectant	encapsulation before freeze-drying (%)	encapsulation after freeze-drying (%)
10% glucose	43.5	20.4
10% sucrose	51.3	30.6
10% mannitol	49.3	39.7
10% trehalose	59.2	40.7

slow cooling: firstly cool down to −18℃ in a refrigerator and then cooled to −65℃.

Table 7.5 Change of encapsulation of ftorafur liposome prepared by ultrasonic irradiance method

lyoprotectant	encapsulation before freeze-drying (%)	encapsulation after freeze-drying (%)
10% glucose	22	9.5
10% sucrose	26.6	13.0
10% mannitol	28.2	18.5
10% trehalose	45.0	35.0

rapid cooling: cool down to -65℃ with a cooling rate of 20℃/min.

2. Measurement of encapsulation of vitamin A liposome

The instruments used to measure the encapsulation of vitamin A liposome are same as those used for ftorafur liposome. The detection wavelength is 328nm and the peak appears at 3.4min.

1) Standard curve

Prepare methanol solution of 10mL with the vitamin A concentration of 1μg/mL, 2μg/mL, 3μg/mL, 4μg/mL, 5μg/mL, 6μg/mL respectively. Pipette 10μL of the prepared solution into the chromatograph through the injection valve. Each solution is measured 7 times in order to obtain the vitamin A chromatogram. The peak area is calculated by use of the chromatogram. Draw a concentration of vitamin A- peak area curve and perform a linear regression analysis to obtain the standard curve for the calculation of vitamin A concentration.

2) Measurement of the content of free ftorafur C_f

Add 500μL liposome suspension with vitamin A into a centrifugal tube and centrifuge at a speed of 3000r/min. Take 6.5μL supernants into a 25mL volumetric flask and then add methanol till the liquid level reach the 25mL mark line. Pipette 10μL prepared solution into the chromatograph through a injection valve. Each solution was measured twice. The content of free ftorafur can be calculated by the standard curve of vitamin A concentration.

3) Measurement of the total amount of vitamin A

Take 6.5μL liposome suspensions with vitamin A into a 25mL volumetric flask and then add methanol till the liquid level reach the 25mL mark line to damage the liposomes. Pipette 10μL prepared solution into the chromatograph through a injection valve. Each solution was measured twice.

The encapsulations of vitamin A liposome measured before and after freeze-drying are showed in Table 7.6.

Table 7.6 Change of encapsulation of vitamin A liposome prepared by ultrosonic irradiance method

lyoprotectant	encapsulation before freeze-drying (%)	encapsulation after freeze-drying (%)
10% glucose	98.8	98.6
10% sucrose	98.5	98.4
10% mannitol	99.1	99.3
10% trehalose	99.2	99.4

rapid cooling: cool down to -65℃ with a cooling rate of 20℃/min.

The following conclusions can be drawn from Table 7.3 to Table 7.6. The encapsulation changes of ftorafur liposome with glucose as lyoprotectant is significant, and the encapsulation before freeze-drying are relatively low. The encapsulation changes of ftorafur liposome before and after freeze-drying with trehalose as lyoprotectant is the smallest, and the encapsulations are high. The encapsulation changes of ftorafur liposome with sucrose and mannitol as lyoprotectants is between the above two lyoprotectants. The high concentration lyoprotectants can reduce the leakage of the freeze-dried liposome pharmaceuticals.

The encapsulations of liposomes formulated by rotating volatile method is greater than and that of ultrasonic irradiance method. The encapsulations of vitamin A formulated by ultrasonic irradiance method is very high, almost 100%, which indicated that the encapsulations of liposomes for fat-soluble drugs is relatively high.

The encapsulations changes of ftorafur liposomes with fast cooling rate is relatively small, which indicated that the encapsulations changes before and after freeze drying is affected by the technical procedures.

8

Disinfection, Sterilization and Validation

8.1 Disinfection and sterilization

8.1.1 Microorganism

1. Microorganisms and Pathogenic microorganisms

The microorganisms are tiny, simple-structure but diverse organisms. According to the structures and compositions, they can be classified as three categories: ①Non-cellular microorganism. They are non-cellular organisms which consist of a nucleic acid and of a proteinic coat, such as virus and subviruse. ②Prokaryotes. Prokaryotes are organisms that lack a cell nucleus and the other organelles found in eukaryotes. They are divided into two groups, the archaea and the bacteria. ③Eukaryotes. Eukaryotes contain organelles such as the cell nucleus, the Golgi apparatus and mitochondria in their cells. The nucleus is an organelle which houses the DNA that makes up a cell's genome. Fungi is a type of eukaryotes.

Most of the microorganisms are vital to humans, animals and plants. Only a small number of microorganisms, known as pathogenic microorganisms, can cause human, animal and plant diseases. Pathogenic microorganisms invade the host, interact with the host's defense systems and result in pathology, a process called infection. According to the source of infection, it can be divided into exogenous infection and endogenous infection. The pathogens that can cause exogenous infection is either from the environment outside the host (such as food) or from the pathogen carriers. The pathogens that can cause endogenous infection is mainly from the patients.

Pathogenicity is the ability of an organism, a pathogen, to produce an infectious disease in another organism. Pathogenicity is determined by the inheritance of micro-organisms species. Virulence is also used to weigh the degree of pathogenicity.

2. Environment for microbial growth

Bacterial growth and reproduction requires sufficient nutrition, appropriate temperature, humidity and suitable environment (such as pH value, etc.). Binary fission is the form of asexual reproduction used by bacteria. Most bacteria are able to divide every 20 minutes under optimum conditions. But for some special bacteria, such as Mycobacterium tuberculosis, it takes 18—20 hours to divide. Under certain conditions, the cytoplasm of some bacteria dehydrates, forming a round or oval body in the internal cell, known as spore. The ability to form spores in the bacteria is determined by the Bacillus spore genes.

A spore is a biological reproductive mechanism, which contains complete nuclein, enzyme systems and the system to synthesize cellular components, that preserves all essential material for bacteria. In an appropriate condition, the spore can grow into a new cell.

Some bacteria do not have spores but have vegetative form, which still has reproduction ability. The spores have very strong resistance to heat, drying, radiation, chemical disinfectants and other physical and chemical factors. Vegetative forms can be killed rapidly in 80℃water, but some spores (such as Botulinum bacteria Bacillus spore) could survive in the boiling water of 100℃ for 3~5 hour[138,139]. Bacterial spores do not cause diseases directly, but when transformed into vegetative forms, they cause diseases because the fast reproduction. High-pressure steam sterilization is the most reliable way to sterilize the contaminated equipments and surgical instruments.

8.1.2 Concept of disinfection and sterilization

1. Disinfection

Disinfection can kill pathogenic micro-organisms, but may not kill the bacterial spores and non-pathogenic micro-organisms. Disinfectant is used for disinfection. At standard concentration, disinfectant can only kill bacterial vegetative forms. To kill spores, higher concentration of disinfectant and longer disinfecting time are needed.

2. Sterilization

Sterilization is a method to kill all microorganisms,such as the pathogenic microorganisms, the non-pathogenic microorganisms, bacterial vegetative forms and spores.

3. Asepsis

Asepsis is the practice to reduce or eliminate contaminants (such as bacteria, viruses, fungi, and parasites) from entering the operative field in surgery or medicine to prevent infection. Sterile rooms, sterile fillings and aseptic operations are basic and necessary technical requirements for the production of drugs, especially for injecting drugs.

4. Pyrogen and its removal

A pyrogen is a substance that induces fever. These can be either internal (endogenous) or external (exogenous) to the body. Exogenous pyrogen ,which is obtained outside the host, are toxins and microorganisms. The pyrogen is basicly bacterial endotoxin (such as lipopolysaccharide), as is the case for drugs. From the aspect of drug-inspection, endotoxin means only pyrogen. Studies have shown that, after entering into the human blood, pyrogen does not directly cause fever and other toxic reaction, but will enable cell (mononucleosis) to release "endogenous pyrogen" which is a mixture of pyrogen and cytokinin(IL -1、 tumor necrosis factor and interferon, etc). The endogenous pyrogen can stimulate the central nervous system (brain cells) to give signal of heating.

To remove pyrogen from objects usually requires dry heat at 180℃ for more than 2 hours or dry heat at 250℃ for more than 30 minutes[138,139].

8.1.3 Basic methods of disinfection and sterilization

1. Physical methods of disinfection and sterilization

The physical methods to kill micro-organisms include heating, electromagnetic radiation, filtration and ultrasonic waves. Heating is the most important disinfection method because high-temperature can lead to microbe enzymes and proteins degenerated or coagulated (structural changes lead to loss of function), thus disrupt their metabolisms to result in the abiosis of microorganisms. Heat disinfection and sterilization can be divided into two major categories : damp heat and dry heat.

(1) Dry heat disinfection and sterilization. Dry heat is to sterilize using high temperature air with relative humidity less than 20%. The heat transfer coefficiency is lower for dry heat sterilization because air is the medium. Vegetative forms can be killed by dry heating at 80—100℃ for 1 hour; while spores can be killed at 160—170℃ for 2 hours [138,139].

a. Combustion: it is a simple, rapid and thorough sterilization method. Its application is limited because of its high destructiveness.

b. Cauterization: it is a way to sterilize directly with flame. It is suitable for the sterilization of vaccination rings and the mouth of the test-tube etc. in micro-organism laboratory.

c. Burning: it is appliable to surgical dressings that infected by some special microorganisms, such as tetanus, gas gangrene, aeruginosa, as well as other disposable items that have been polluted, such as paper sewage and garbage. All of the above should be burned down in incinerators.

d. Hot air oven: it sterilizes by hot air in the oven. The hot air in the oven forms a even temperature flow by convection to heat the objects directly. Vegetative forms can be killed in the dry air at 80—100℃ for 1 hour; spores can be killed at 160—170℃ for 2 hours. Dry roast method can sterilize glasses, porcelain, glass syringes and so on.

(2) Moist heat disinfection and sterilization. In moist heat disinfection and sterilization, heat is transfer by the combination of air and water vapor, which is more effective. Damp heat sterilization required lower temperature and shorter time compared with dry heat sterilization method.

a. Pasteurization: Pasteurization is usually used in dairy production to serilize at 60℃ for 30 minutes in order to kill pathogenic microorganisms and to protect the nutrients. In biomedical production (such as human serum albumin) process , Pasteurization sterilization requires 60℃ for 10 hours to kill virus.

b. Boiling: Under atmospheric pressure, the boiling water at 100℃ can kill bacterial vegetative form in 5—10 minutes and spores in 1—2 hours. The boiling point of aqueous solution with 2% sodium carbonate solution is 105℃, which can both improve sterilization ability and remove dirts.

c. High-pressure steam: An autoclave is a pressurized device designed to heat aqueous solutions above their boiling point at normal atmospheric pressure to achieve sterilization. Elevated temperature and pressure are used to sterilize in high-pressure steam method. It is applicable to items that can resist high temperature, high pressure and moisture, such as surgical dressing, surgical instruments, filter for sterile, rubber plug, aluminum-plastic compound cover and bacterial culture medium, etc.

(3) Ultraviolet disinfection. Ultraviolet ray of 200—300nm wavelength can be used for disinfection. Because ultraviolet ray has low-energy and cannot penetrate through a glass、dust、paper and solid substances, it can be only used to disinfect air in the operating rooms, infectious diseases houses and

microbiology laboratories, or used to disinfect the surface of items that are not resist to heat . Ultraviolet ray of these wavelengths is harmful to human skin and eye.

(4) Ionizing radiation sterilization. The high-energy electron beams produced by radioisotope source or linear accelerator can be used for sterilization. Ionizing radiation is also called "cold sterilization" for it can be performed at room-temperature. It is specially suitable for disposable medical equipments, packaged equipments, and precision medical devices.

(5) Filtration. Filtration realized stelization by physical seperation to remove bacteria in liquid or air. The filters contain tiny pores through which only liquid or gas can flow. The larger particles (such as bacteria) will be blocked as their size is greater than the diameter of the pores. Filtration is mainly used to sterilize some of the non heat-resistance biological agents.

High efficiency particulate air filter is a kind of modern equipment which is usually adopted in pharmaceutical companies or in hospitals for the purpose of air purification. It uses filters with the pores' diameter less than 0.22μm for stelization. This method is widely used in the aseptic pharmaceutical filling rooms, hospital operation rooms and so on.

(6) Ultrasonic wave. 20—200 kHz ultrasonic waves can break bacterial cell, so it is applied for disinfection purpose. Ultrasonic washer can be used to clean hands and syringers .

2. Chemical disinfection and sterilization

When chemical substances penetrate into bacteria, they will denature bacterial protein, decrease the activity of bacterial enzyme, inhibit bacterial growth and metabolism, or damage the structure of cell membrane, change membrane permeability. So chemical substances, called chemical sanitizer in this case, can be used for disinfection and sterilization.

(1) According to the mechanisms of disinfection, chemical disinfectants can be divided into the following categories:

a. Coagulate and denature bacterial protein, such as phenols (high concentration), alcohols, heavy-metal salts (high concentration), acids and alkalis, aldehydes.

b. Interfere with bacterial enzyme system and metabolism. such as certain oxidants, heavy-metal salts (low concentration). They unite with bacterial SH-base and cause loss of their enzyme activity.

c. Damage cell membranes of bacteria, such as phenols (low concentration), surfactants. They can lower surface tension of bacterial cell membranes and increase membrane permeability.

(2) Methods for disinfection and sterilization with chemical disinfectants

a. Immersion: Disinfection and sterilization are realized by immersing items in disinfectants that can kill most type of bacteria , less corrosive and water-soluble.

b. Wipe: Disinfection and sterilization was realized by wiping the surface of objects with water-soluble and strong-penetrating disinfectants.

c. Fumigation: Vaporize the disinfectants by heating or oxidants for disinfection. It is applicable to disinfect indoor goods and air; or expensive precise equipments and items that can not be steamed, boiled, soaked. Clean rooms are usually fumigated with formaldehyde or peracetic acid, etc.

d. Spray: Spray the disinfectants into the sterilized space or objects by sprayer or aerosol sprayer for disinfection.

e. Ethylene oxide sterilization: Charge a closed container with ethylene oxide for the purpose of disinfection and sterilization. Ethylene oxide can kill most types of bacteria, such as spores, fungi and viruses. Ethylene oxide is volatile and has strong penetration ability. It is harmless for most of the materials, especially applicable to those objects that can not intolerance high temperature and heat, such as precise instruments, electronic equipments etc.

(3) Characters and Efficiency of the disinfectants

a. High-level disinfectant: Can kill most types of bacteria can be used in many ways. High-level disinfectants include ethylene oxide, peracetic acid, formaldehyde, glutaraldehyde, disinfectant chlorine (such as bleaching powder、 three-in-one、 sodium hypochlorite、 superior chlorine, etc.). High-level disinfectants should be used soon after formulated as they are chemically unstable.

b. Middle-level disinfectants:They have good solubility and long shelf life. They are chemically stable but can not be used as sterilizing agents. Middle-level disinfectants include iodine, iodine tincture ethanol, lysol and potassium permanganate etc.

c. Low-level disinfectant: They are chemically stable, odorless, irritationless and have long shelf life. They can kill a few types of bacteria and can only inhibite but not kill spores. Low-level disinfectants include quaternary ammonium salts (such as Dimethyl benzyl lauryl ammonium bromide、 Domiphen Bromide、 Myristylpicoline Bromide).

8.2 Validation

8.2.1 Importance of validation

1. Concepts and importance of validation [140—146]

In the medical device, pharmaceutical and biotechnology manufacturing industries, validation refers to establishing documented evidence that a process or system, when operated within established parameters, can perform effectively and reproducibly to produce a medicinal product meeting its pre-determined specifications and quality attributes.

Documents of validation should at least include the followings: *General plan of validation*, *Plan of validation*, *Validation report* and *Summary of validation*. *Validation report* should include the initial installation plan, the final approval document, the confirmed installation plan, the confirmed run program, the product validation, the cleaning validation, the examination validation and the authorized certificate and so on. Up to now, there is no universal model for GMP validation. The validation plan is usually developed according to the specific requirements of GMP as well as the situation of different pharmaceutical. In particular, validation parameters is an essential but hard to set.The parameters should be not only enforceable but also representative, if necessary, they can be applied to some challenging experiments. Despite validation is a manpower and resource costing work, all of the drug manufacturers pay great attention to it for three reasons:

(1) Governmental regulation: Both FDA (American Food and Drug Administration) and SFDA (Chinese State Food and Drug Administration) has mandatory requirements for pharmaceutical manufacturers to enforce the validation. They emphasize that the combination of GMP and validation can guarantee quality of products.

(2) Guarantee of theproduct quality: The validated technical procedures will guarantee the quality of product. It is an essential part of quality assurance system, which helps us understand and control technical procedures.

(3) Reduction of initial costs: A properly validated and controlled process will yield little scrap or rework, resulting in increased output. Consistent conformance to specifications is likely to result in fewer complaints and recalls. The original motivation of validation is to reduce costs and ensure product quality.

2. Classification of validation

(1) Prospective validation includes those considerations that should be made before an entirely new product is introduced by a firm or when there is a change in the manufacturing process which may affect the product's characteristics, such as uniformity and identity. The purpose of prospective validation is to verify whether the changes of operating parameters meet the changes of independent freezing-drying variables and product quality.

(2) Concurrent validation is conducted with the intention of ultimately distributing product manufactured during the validation study. Concurrent validation is feasible when nondestructive testing is adequate to verify that products meet predetermined specifications and quality attributes. If concurrent validation is being conducted as the initial validation of a new process or a process which has been modified, product should be withheld from distribution until all data and results of the validation study have been reviewed, and it has been determined that the process has been adequately validated.

(3) Retrospective validation is the validation of a process based on accumulated historical production, testing, control, and other information for a product already in production and distribution. This type of validation makes use of historical data and information which may be found in batch records, production log books, lot records, control charts, test and inspection results, customer complaints or lack of complaints, field failure reports, service reports, and audit reports. Historical data must contain enough information to provide an in-depth picture of how the process has been operating and whether the product has consistently met its specifications. Retrospective validation may not be feasible if all the appropriate data was not collected, or appropriate data was not collected in a manner which allows adequate analysis.

If historical data is determined to be adequate and representative, an analysis can be conducted to determine whether the process has been operating in a state of control and has consistently produced product which meets its predetermined specifications and quality attributes. The analysis must be documented.

After a validated process has been operating for some time, retrospective validation can be successfully used to confirm continued validation of that process if no significant changes have been made to the process, components, or raw materials.

Statistical process control is a valuable tool for generating the type of data needed for retrospective analysis to revalidate a process and show that it continues to operate in a state of control.

(4) As long as the process operates in a state of control and no changes have been made to the process or output product, the process does not have to be revalidated. Whether the process is operating in a state of control is determined by analyzing day-to-day process control data and any finished device testing data for conformance with specifications and for variability. When changes or process deviations occur, the process must be reviewed and evaluated, and revalidation must be performed where appropriate.

(5) Process validation: Process validation is establishing documented evidence which provides a high degree of assurance that a specific process will consistently produce a product meeting its pre-determined specifications and quality characteristics.. It has two connotations:

a. It proves that a process meets its pre-determined specifications, in other word, the process is under controll.

b. It determines the variables and the conditions to accepte these ariables in a process and establish a corresponding control plan between them.

8.2.2 Developmental trend of validation

1. Developmental trend of concept of validation

At present, quality management of pharmaceutical is expanding its supervising area in the aspect of validation. For example, the concept of validation is being applied to providers of raw materials. Modern pharmaceutical companies not only inspect raw materials, but also evaluate the qualification of their suppliers. The validation to the raw materials and the suplliers is continued through the whole process of production.

Validation also is used in the research process of drugs. FDA and SFDA asks the company to provide validation documents when they apply for the certification of a new drug. The process not been validated cannot be written in the application documents. Research department must provide the detail documents for production process and is responsible for directing the production department to validate 1—3 batches of approved production process.

2. More explicit goal of validation

The goal of peocess validation is to prove the reliability and repeatability of the process. To confirm the performance is to confirm what result will be achieved under pre-determined conditions. It should be emphasized that process validation and process optimization are two entirely different concepts: process validation is not to optimize conditions of process but to prove the reliability and repeatability of the technology; while process optimization is try to get the best technical parameters in product research and development stage. The new product development stage should be very slight, including the documents for laboratory test and middle test, moreover, process optimization of large-scale production should be complete before new product put into production. Before producing the new product, production department should confirm the reliability and repeatability of the technology, perfect the regulation train personnel, and then start the conventional production.

8.3 Cleaning-in-place, sterilization-in-place and validation

8.3.1 Cleaning-in-place and confirmation of freeze-drying system

The front part of a freeze-dryer (door, operation surface etc.) is located in the cleanroom. It cannot be used until suffocated by formaldehyde gas and reached the standard of class 100 cleanroom.

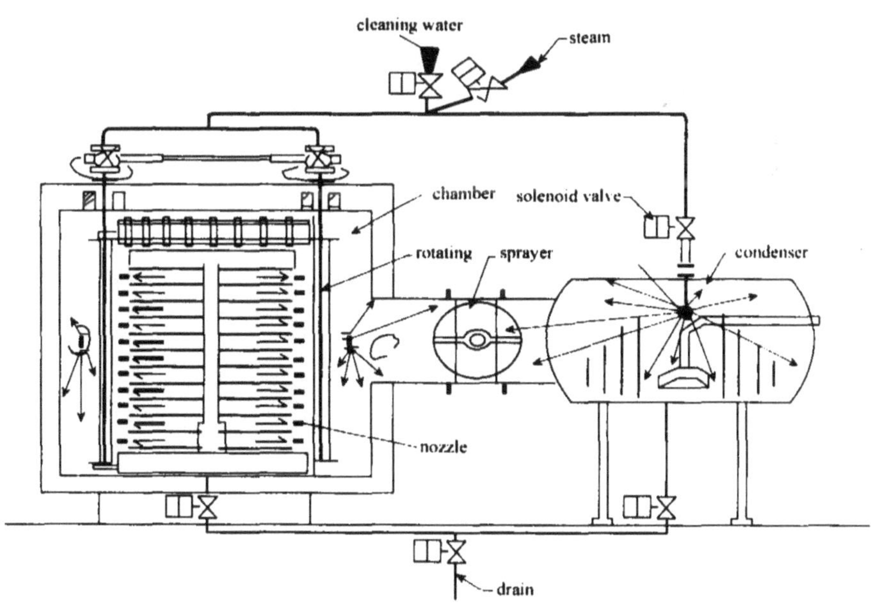

Fig.8.1 Diagram of CIP of freeze-dryer

1. Cleaning in place(CIP)

Cleaning in place (Fig.8.1): generally, washing with reverse osmosis (RO) water for 5 min, followed by water for injection (WFI) for 5 min.

2. Validation of Cleanness

After cleaning, check the cleaned surface by wiping test to confirm if there are dirts, active agents or bacteria on the surface. Comparing the results before checking with that after checking to ensure the cleaning is conform to GMP standard.

8.3.2 Sterilization in place and confirmation of freeze-drying system

1. Sterilization in place (SIP)

Because drugs produced by freeze-drying are aseptic, and the drugs are exposed to the whole freeze-drying system, freeze-dryer's chamber and the condenser should be regarded as a cleanroom. The srerilization issue must be considered when designing the freeze-dryer. At present, saturated steam sterilization method is widely used in freeze-dryer. 316L stainless steel is used to manufacture freeze-dryer's chamber for its good corrosion resistance ability (Fig.8.2).

Moist heat sterilization use the saturated steam to sterilization. It has the charcter of high efficiency and low price. The freeze dryer is sterilized using saturated steam at 121℃ for 30min.

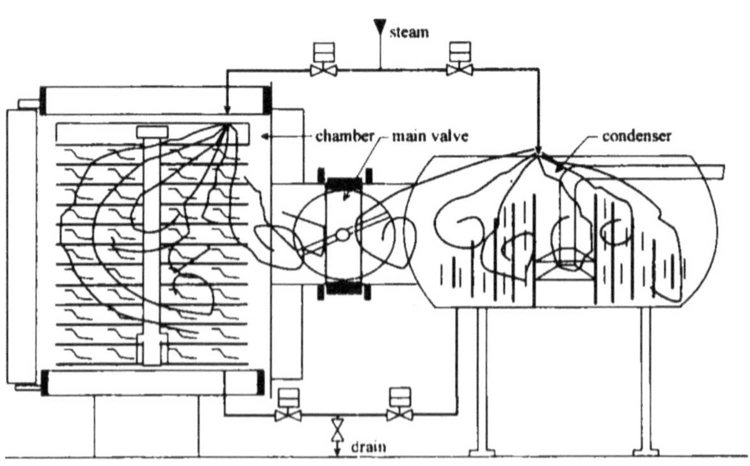

Fig.8.2 Diagram of SIP of freeze-dryer

2. Confirmation of SIP

(1) Thermocouple verification: The thermocouple used to indicate the sterilization temperature must be verified before and after the SIP. The difference between a standard thermometer and the thermocouple must be less ±0.5℃.

(2) Temperature profile test: Generally the temperature of the thermocouple in the chamber can not reflect the temperature profile in freeze-dryer. More than 10 thermocouples should be placed in the chamber when performing SIP test. Repeat the SIP three times at 121℃ for 30min and record the position of the thermocouple given the lowest temperature. The difference between the lowest temperature and the mean temperature must be less than ±2.5℃.

(3) Biological indicator challenge test: Put 10—20 sealed ampoules with 10^6 Bacillus Sterothermophilus each in the fixed position of the shelves, put the verified thermocouples beside the ampoules and start the SIP test. According to the USP, the Bacillus Sterothermophilus can be killed at 121℃ in at least 30 min . After sterilization, test to the ampoule according to "China Biological production Regulations" (2000). If the result is negative, it indicates sterilization meets the requirment of standard. If the result is positive, adjust sterilizing procedure, extend the sterilizing time and repeat the SIP test.

8.4 Process validation of freeze-drying

8.4.1 Validation protocol

1. General validation protocol

The validation protocols are the documents with tests and instructions that are executed for a process, equipment validation or analytical method validation etc.

A complete process validation includes the fllowing: the public project confirmation and the manufacture environment confirmation, the mechanical device's installment and the operating condition confirmation, the operating procedure confirmation, the producing technique and the product confirmation. The detailed validation protocols should also include the operating procedures for each validation test.

2. Process validation document of freeze-drying

The freeze-drying of agents include most of the technical processs for many other types injection production, such as preparation, filtration, washing, sterilization, package and so on. The process validation document of freeze-drying include the illustration of equipment operating principle, installation condition, operating condition, principle of control system as well as the related public project and service facility. The validation protocol should include Standard Operating Procedure (SOP)、 data acquisition method and the data processing method. The document should also include the test condition and result for the equipments, the freezing rate with/without load, the lowest temperature, the precision of the shelf temperature、 the pressure of the freeze-dryer, the precision of the vacuum, the condition to get the limited vacuum, the efficiency of sterilization. In this chapter we only illustrate some basic concepts of process validation of freeze-drying form engineering point.

8.4.2 Freeze-drying system and its accessory

1. Freeze-drying system

A freeze-drying system composes of chamber, the condenser, the heat-exchanger, the refrigeration system, the vacuum system and the controll system. The equipments and devices in the freeze-drying system, including their detailed technology document, assembly blueprint, are important for validation protocols. These materials should be archived as part of the validation documents.

2. Public system and the accessories

Freeze-drying system also needs a suitable public system and accessories, such as electricity, compressed air, cooling water, water for injection (WFI), steam supply system. The performance of public system will affect the safety and performance of the freeze-dryer. For example, shortage of cooling water will result in the increase of condensing pressure and consiquent reduction of refrigeration capacity, as a result, the sublimation rate will be decreased and sublimation time will be extended. If the refrigeration capacity is too small, the vacuum will be lose controll and the frozen agents will melt.

In general, the capacity of the public system is greater than the requirements of the freeze drying system to meet the change of the freeze drying load. All of the above should be archived as part of the validation documents.

3. Verification and maintaince of metering devices

In the freeze-drying process the temperature and the pressure must be monitored. In order to obtain accurate and reliable data, the metering devices must be adjusted before and after validation besides the regular verification.

8.4.3 Performance and parameter validation of freeze-drying system

The confirmation of performance parameters take a important part in the validation of freeze drying process. The design and performance of the freeze dryer should be validated before the validation of technical process of agent freeze drying. Only when the parameters and the reliability of the whole freeze drying system match the design values, the pre-determined technical process of agent freeze drying can be obtained.

The characters of the freeze-drying equipment: the product temperature always changes between – 40℃to 50℃.the temperature of the condenser (cold trap) should always be maintained between –70℃ to 50℃. refrigeration load fuctuation greatly. Large refrigeration capacity is required during the freeezing of the agents, but small refrigeration capacity can maintain the freeze drying process during the primary and the second drying process. So the refrigeration system always operates with small load and low efficiency.

1. Validation of the main freezer's performance

The function of the main freezer is to freeze the agent below its triple point temperature or the glass transition temperature, and provides cold in the drying process. The main freezer's performance need to be validated at the condition of zero load and water load.

(1) The validation of the cooling capability at zero loaded.

At zero load, the validation of the temperature decrease rate of the vacuum condenser or the shelf is performed after the installation (installment confirmation) or maintaince (re-validation). Before this validation, make sure that there is no leakage of the accessories of the refrigeration system; the charge of the refrigerant is in the standard amount; the operating pressure of the freezer and the temperature of the cooling water is normal; all of the indicated temperatures in the control pannel are normal. Then when the main freezer is operated at full load, its performance should reach the following:

a. The suggested standard for the shelf temperature: the time cannot exceed 70 min. when the shelf temperature decreases from 10℃ to –40℃.

b. The suggested standard for the average decrease rate of shelf temperature: the time cannot exceed 90 min. when the shelf temperature decreases from 10℃ to –50℃, that is, the average decrease rate should be greater than 1.5℃/min.

c. The suggested standard for the temperature of vacuum condenser: the vacuum condenser temperature should be below –70℃。

(2) The validation of the cooling capability at water loaded. After validating the cooling capability at zero load, put same distilled water in the shelf (the load od the water is equivelent to the maxmun load of the agents to be freeze dried), and start the main freezer. The suggested parameters should be:

a. The temperature decrease rate of the shelf: the time should be less than 100min for the temperature of the water to decreases from 10℃ to -35℃.

b. The time should be less than 120min for the temperature of the water to decreases from 10℃ to –45℃.

c. The vacuum condenser's temperature is kept below –55℃. If the there is more organic compounds in the agents, the vacuum condenser's temperature is kept below –65℃.

(3) Validation of vacuum condenser's temperature precision.

At the different stages of the freeze-drying process, the cooling load charges greatly. Although the vacuum condenser's temperature is controlled below the water's saturation temperature corresponding to the vacuum by load-adjusting device, the temperature in the vacuum condenser will fluctuate because of the discontinuous running of the load-adjusting device, and this fluctuation will influence directly the pressure of the vacuum system. So it is important to make sure if the temperature fluctuation of the vacuum condenser meet the vacuum requirement of the agents. Generally the temperature fluctuation should be controlled in ±3℃.

(4) The validation of the vacuum condenser's maximum water-condensation ability.

The vacuum condenser's maximum water-condensation ability is described in the mannual of the freeze dryer, which is obtained at standard condition. It is necessary to confirm if the freeze-dryer's maximum water-condensation ability can be obtained because the users provide the freeze dryer with different working condition. In principle, the vacuum condenser's maximum water-condensation ability can be validated by overload test. Put excess water in flat tray, stop the freeze dryer when it is full of ice in the vacuum condenser, weigh the remaining water in the tray and calculate the difference of the distilled water before and after the operation of the freeze drier.

2. The validation of the freezer's temperature- control ability

In the freeze-drying process, the main functions of the control freeze drier: exactly compensate the fluctuation of the agents' temperature in the sublimation drying process. The temperature control ability mainly depend on the match (±1.0 ℃) between the refrigeration capacity and the load changes.

Heating capability validation: it needs to proride heat to the freeze drying system in order to provide phase change heat to the sublimation and the desorption energy in second drying process. These heats usually provided by electric heater.

The validation of the heating system:

(1) The heating rate validation under zero load. Recommend criterion: under zero load condition, electric heater run at full load, the heating rate should be greater than 25 ℃/h. and the final temperature can reach the pre-determined temperature in the mannual, which is uauslly 50℃.

(2) The heating rate validation of under full load. Recommend criterion: under the full water load condition, electric heater run at full load, the heating rate should be greater than 20℃/h.

The validation of the heating rate control. Recommend criterion: The heating temperature should be controlled at ±1.0℃ more or less the required temperature.

3. The validation of the vacuum system's performance

(1) Vacuometer and its verification. The vapor partial pressure of the chamber must be below the triple point of water. The vacuum pumps in freeze-drying system are assembled based on the vacuum of 1×10^{5}—1.3×10^{-1}Pa for low vacuum and 1.3×10^{2}—1.3×10^{-1}Pa for middle vacuum. The gas in low vacuum is viscous flow, the remaining gas are air and vapor. The oil-sealed rotary vacuum pump is uauslly used to exhaust the gas in low vacuum. In middle vacuum, the remaining gas is mainly water vapor. The combination of Roots pump or jet pump with rotary pump are used to exhaust the gas in middle vacuum. The thermal conductivity vacuum gauges (Pirani gauges) is used to measure the vacuum. The verification of this type gauges should be made by the qualified units.

(2) The validation of the vacuum pump's exhaust rate. A typical vacuum system of a freeze-dryer is usually compose of two oil-sealed rotary vacuum pumps and one Roots pump. In exhaust rate validation test, a known-volume sealed box (for example the freeze-dryer's volume, including chamber volume, vacuum condenser volume, vacuum pipeline volume etc.) is used. Record the time exhausting the volume from atmospheric pressure to 133.3 Pa and then calculate the actual exhaust rate (L/min).

(3) The validation of vacuum system's performance.

a. The recommended standard of the elementary vacuum pump's performance: under zero load condition and charged with new oil, The pressure should decreased form 105Pa to 6.7Pa in 10min (from the start of vacuum pump).

b. The recommended standard of the elementary vacuum pump + Roots pump: under zero load condition, the vacuum should be 1.33Pa in 5min (from the start of Roots pump).

c. The recommended standard of the limited vacuum under zero load: under zero load condition the pressure is decreased below 13.3Pa in 20min, and 1.33Pa in 6 hours.

(4) The validation test of vacuum system leakage. Vacuum system leakage means that exterior air enters into the freeze-dryer's vacuum system. The total leakage amount is used to evaluate the seal of vacuum system. The acceptable standard: $\Delta L<200$ L·μmHg/s.

(5) The validation of vacuum control. The purpose to control vacuum in the chamber is different between the primary drying and secondary drying. In primary drying, the aim to keep a constant pressure is to make the ice crystal sublimat fast and keep the materiel in structure. The aim in the secondary drying is to increase heat exchange between agents and shelf to remove the remaining water. The vacuum in secondary drying is more important because the moisture content of final freeze-dried agents will determine its shelf life. There are two methods to control vacuum: induced gas and exhaust rate. The first one is maintain the pressure of the chamber by induceing suitable gas(air or nitrogen) into the chamber. The second one is by adjusting the exhasut valve to maintain the pressure. No matter which method is used, the control precision of the vacuum should be validated by test.

4. Others

At present, the freeze-drying system is equipted with online monitor system and intellectual control system, which help set up technical procedures according to the requirement. It can greatly improve the control precision of the technological parameter and the agents' quality. The supporting software of the online monitor system and intellectual control system must be validated.

Besides the freeze dryer, the equipments in the producing technicals of freeze dried agents include washing machines, tunnel sterilization, Semi-filler and rubber plug, rolling cover, steam sterilization and filtration system. These equipments are required to meet the standards of GMP in design, materials and operation.

8.5 Validation of technical process and the evaluation of validation

8.5.1 Technical process of freeze-drying

The modern freeze-drying is mainly used in the production of biological agents and antibiotics with unstable properties. A typical lyophilized process is described in Fig.8.3.

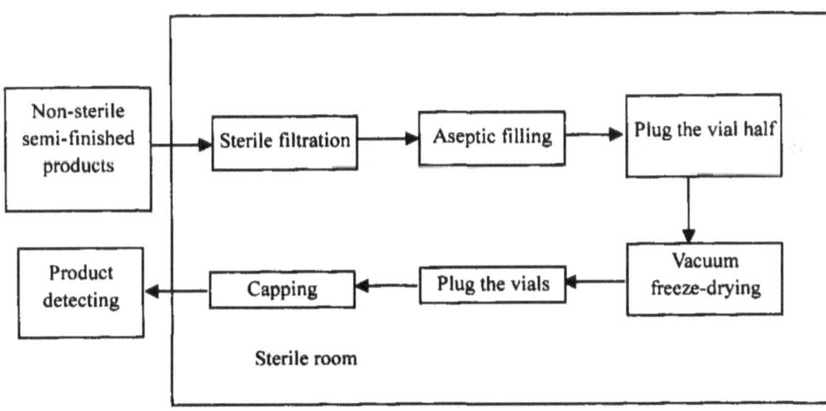

Fig.8.3 Flow chart of a typicl lyophilized process

The validation of technical process includes the prospective validation of the new product and retrospective validation of the old products. Prospective validation includes those considerations that should be made before an entirely new product is introduced by a firm or when there is a change in the manufacturing process which may affect the product's characteristics, such as uniformity and identity. In some cases a product may have been on the market without sufficient premarket process validation. In these cases, it may be possible to validate, in some measure, the adequacy of the process by examination of accumulated test data on the product and records of the manufacturing procedures used, which is called retrospective validation. Retrospective validation can also be useful to augment initial premarket prospective validation for new products or changed processes. In such cases, preliminary prospective validation should have been sufficient to warrant product marketing. As additional data is gathered on production lots, such data can be used to build confidence in the adequacy of the process. Conversely, such data may indicate a declining confidence in the process and a commensurate need for corrective changes. Test data may be useful only if the methods and results are adequately specific. As with prospective validation, it may be insufficient to assess the process solely on the basis of lot by lot conformance to specifications if test results are merely expressed in terms of pass/fail. Specific results, on the other hand, can be statistically analyzed and a determination can be made of what variance in data can be expected. It is important to maintain records which describe the operating characteristics of the process, e.g., time, temperature, humidity, and equipment settings. Whenever test data are used to demonstrate conformance to specifications, it is important that the test methodology be qualified to assure that test results are objective and accurate.

There are many parameters in the lyophilization technical process, but it is not necessary to invole all parameters in the validation. It is very important to chose some significant parameters for validation, especially those that has significant impact on the quality of the final products. The weak links of the equipment can be found by this valiation test.

The technical process validation documents should include analysis and evaluation of the data of batch production process and the quality of final product. The technical process parameters include freeze-drying time, temperature, vacuum and so on. It is necessary to satistically analyze the relationship between controlled parameters and quality of final products, the deviation between the test results of final product and the pre-determined data, to ensure the parameters set in the validation plan meet the requirements of standard. Main contents of the technical process validation will be described as follows.

1. Process water and freezing rate

1) Process water

In the manufacturing process of lyophilized preparation,the water or solvent used to compond liquid medicine is known as process water. Process water, which takes a major portion of liquid medicine, is helpful to improve the accuracy of the dose of medicine. However, the composition of process water will affect the structure of ice crystals and the rehydration performance of the preparation. The ice crystals should be uniform in the frozen structure of the liquid medicine. It is not an unideal structure to form a hard shell or a thin layer on the surface of the preparation, which will resist the flow of vapor in the drying process and reduce the drying rate.

There is a supercooling in the freezing process of the liquid medicine.The supercooling has great influence on the structure of the ice. When supercooling is more than 10℃ lower than freezing point, the cells and protein's activity will be damaged and the sublimation channel become small to result in the decrease of drying rate.

The supercooling is especially affected by other components (sugar, starch; sodium and potassium and excipients) in the process water. These components have great effect on thermophysical properties of liquid medicine and may induce the change of pH value, rehydration time and solution turbidity. The validation test of process water is very important.

2) Freezing rate

Generally, the capacity of freeze-drying equipment is fixed at standard condition. However, the capacity will be affected by the load of the products, the climate and so on. The change of capacity will affect the freezing rate. Therefore, the freezing rate of the products in freezing stage should be validated.

Generally, large ice crystals are desired to form to maintain the unimpeded sublimation channel to increase the sublimation rate. But too large ice crystals will affect the uniform of ice, which decrease the sublimation rate. For some polymeric drug with non-regular network structure,fast freezing will fix their structure rapidly and the organic subsatnce will voletil rapidly. Therefore,the opetimal freezing rate for lyophilized preparation is depended on the its characteristics. The freezing rate for different preparations need to be validated to conform it with the requirement of the technical process.

2. Preparation temperature and drying time

Generally, preparation temperature is measured by temperature sensors which are directly placed in glass vials or in the trays. Though the preparation temperature can be measured directly, it is controlled indirectly by temperature change of the shelf. Therefore, validation test should determine the relationship between the pre-determined shelf temperature and the measured preparation temperature in freeze drying process to verify the temperature change can meet the technical requirements.

In primary drying process, lower preparation temperature can guarante the quality of the preparation but cost longer time for drying. While higher preparation temperature will reduce its quality. Therefore, the preparation temperature should be validated for the purpose of optimizing the technical parameters.

The validation of the preparation temperature is suggested to be done as follows:

1) The preparation temperature in sublimation drying process

Theoretically, the heat transferred to the preparation should be equal to the heat required fore sublimation. If there is not enough heat provided to the preparation, the temperature of the sublimation front will drop, and then the sublimation rate will decrease. If excess heat is transferred to the preparation, the preparation temperature will increase and result in melt of the preparation.

The optimum preparation temperature is the one that can keep the fastest sublimation rate and the frozen of freezing layer melt. In the sublimation drying process, the validation test for the effect of the preparation temperature and shelf temperature on the quality of products can be carried out together.

2) The determination of the preparation temperature and end of drying in the secondary drying process

The validation in the secondary drying process is to mainly verify whether the preparation temperature elevation will affect the preparation quality. If the preparation is exposed to higher temperature for longer time, the preparation may discolor. The preparation temperature rises gradually, and equals to the shelf temperature at last. If there is not precise method to determine whether the temperature satisfy the requirements, it is difficult to determine whether drying is complete. Excessive drying will cause waste of energy and time. On the other hand, if the preparation temperature is too low, the water content of the final products will be greater than the requirements of the standard.

It is also need to verify whether the method of determine the end of drying is right. At present, residual gas method is used to judge the end of the freeze drying process. The procedure is: disconnect the freeze drying chamber and vacuum pump, observe the change of vacuum in the chamber. For vapor, the drying process is finished when the change speed of pressure is less than 5Pa/3min. This method should be verified by analysis of the water residues in the preparation.

3) Effect of vacuum control precision on preparation

In the sublimation process, gas is sometimes charged into the chamber to enhance convective heat transfer in the compensation of pressure increase.

Therefore, the optimum amount of gas should be charged to get a balance between heat transfer rate and vacuum. The appropriate vacuum control precision is usually determined by the technical process of freeze drying. When the thickness of ice is becoming thinner, vacuum control precision become higher. The conclusion of this validation test is evaluated by the appearance analysis and the quality of the products.

4) Validation of vacuum condenser temperature

Vacuum condenser is used to condense vapor. In the sublimation drying process, the working temperature of the vacuum condenser is –75℃—–50℃ and the corresponding preparation temperature is –35℃—–10℃. When the temperature is –76℃, the saturated vapor pressure of the ice is about 0.1Pa. The drying rate will not change much when the temperature of the vacuum condenser is below –76℃. The working temperature of the vacuum condenser is usually kept at about –60℃.

The validation of the working temperature of the vacuum condenser is to calculate the surface temperature of vacuum condenser and compare to the measured temperature, in order to justify whether vacuum condenser is working at the right temperature. The vacuum condenser should work at the lower temperature if there is organic solvent (alcohol, methanol and so on) in the preparation solution.

8.5.2 Evaluation and confirmation of the validation results of the freeze dried products

Products validation in the freeze drying technical validation is a trial production under special monitoring conditions. It is a continuation of installation confirmation and operation confirmation for various equipment and accessories. Its main purpose is to perform a systematic validation for the reliability and reproducibility of the technical conditions by trial production.

The validation of the freeze dried products usually includes the following.

1. Confirmation of freeze drying technical parameters

This requires to confirm the parameters control according to the manufacturing process procedures. Usually, the research departments have a set of complete technical parameters and control programs before carrying on trial production. However, because the practical conditions, the equipments, and public facilities can't completely meet with the technical parameters and control program provided by the research departments, it is necessary to design experiment in the trial production to determine the freeze drying processures. It is also required to repeat three times of the test and performing comprehensive quality inspection (including visual examination and moisture test) of the samples to ensure the applicability and reproducibility of the freeze drying procedure.

2. Confirmation of the quality of lyophilized preparation

1) Appearance and redydration

The appearance of freeze dried products depends on the freezing condition and distribution of ice crystals. The color of the freeze dried products is related to the preparation's thermal stability ,temperature and drying time. The structure and color will be affected if the freeze drying system breakdown in the freeze drying process, especially in the primary drying process. The resulted semi products must be heat dry in the secondary drying process. And the final produts has rough surface, honeycomb-like cavities and yellow color.

Moreover, the poor chemical or physical treatment to the inner surface of the chamber and the big fluctuation of vacuum will lead to defects such as inconsistent shrinkage and concavo-convex top, preparation lamination, uneven color of each layer and so on. The final product of lyophilized preparation should be perform rehydration test according to "The Chinese Pharmacopoeia" (2000 edition) or "China Biological preparations Regulations" (2000 edition and 2002 supplementation). The rehydration time, clarity of solution, the particle size and so on must conform to the requirements of the regulations.

2) Moisture

The validation of water content of the final lyophilized preparation is to measure the water content and moisture uniformity under the same technical process. The deviation of moisture in the final products is caused by different shelf temperature. In the validation process, through the statistical analysis of water content to confirm whether the freeze-drying procedure is right. This validation may be performed generally in the expanded production experiment. Generally, the moisture of lyophilized preparation must be lower than 3%.

3) Potency

Potency means the activity of biological preparation. Potency will be reduced in the freeze drying process. In the secondary drying process, the preparation was exposed longer time at 30—35℃, which cause great potency reduction.

For lyophilized biological preparation, the potency of final products is 80%—140% of labelled amount on drug bottles.

4) Product stability testing

Stability testing data of 3 consecutive batches of products is provided in the declaration documents of drug approval. In recent years, products stability testing has been used in products validation of lyophilized preparation.

Accelerated testing: Under the condition that the temperature is (40±2)℃, and the relative humidity is (75±5)%, products was stored for 6 months and was tested at the 1, 2, 3, 6 month intervals. For drugs sensitive to temperature, the condition is temperature of (25±2)℃, the relative humidity of (60±10) %.

Long-term testing: Under the condition that the temperature is (25±2)℃, and the relative humidity is (60±10)%, products was stored for 36 months and was tested at the 0, 3, 6, 9, 12, 18, 24, 36 month intervals. Drugs sensitive to temperature can be stored at (6±2)℃ for 12 months and tested according to the requirement above.

3. Evaluation and confirmation of validation results

Compare its validation results with acceptance criterion to form conclusive evaluation after the validation test according to the various validation plan.

The acceptance criterion is established based on verification environment. For example, the acceptance criterion is defferent between the new equipment and the maintained equipment. For a new equipment, its acceptance criterion is based on the mannual. For the maintained equipment, only the maintaied part need to be validated and retropective verification is needed for the whole system. If the product variety and the specification changes, retropective verification of the equipment technical parameters is needed.

1) Factors influencing the properties of lyophilized preparation

If freeze drying technical process and the equipment maintain stable, then the test results of the final preparation can be used for evaluation of process validation. Both the quantity and contents of the items in the quality standard should be cosidered in the validation process.

It is more important to consider the importance of preparation characteristic than the quantity of characteristics. For example,when comparing the stability and color of preparation, the stability has much greater influence than the color on the preparation quality. But it doesn't mean that the color of the preparation can be ignored. It just demonstrates that, for different preparation, the influence of each characteristic to the preparation is different.

The confidence level of validation for freeze drying technical process: there are two choice in setting up the validation of technical process. One is listing all the characteristics items of the lyophilization preparation; the other is only considering the items which have great influence on the preparation properties. Some properties, such as the color, are difficult to perform validation because there are subjective factors. Therefore, color is less importance than other factors in validation. Of course, the confidence level of the validation results will also be affected to a certain extent.

2) Processing methods for unmatching acceptance criterion

If validation results do not match acceptance criterion, it is necessary to investigate the reasons. According to the findings, the initial accepted standards can be revised with extremely carefulness. Because the acceptance criterion are generally legal requirement. The target value cannot be obtained by results even if there is no normative acceptance criterion.

8.6 Daily monitoring, revalidation and retrospective validation

8.6.1 Daily monitoring of freeze drying technical process

The public facility as water, electricity, air, and vapor may not be as stable as the requirement in very long period of freeze drying. In addition, the unstable operation of the freeze-drying equipment may induced by wear and tear of parts. The standard operating procedures will not cover all the factors in the technical process. As a result, it is necessary to strengthen management and master control of these change factors in the daily production process in order to obtain the expected results of the design, ensure that the operation of all conditions and equipment of freeze drying technical process maintain the conditions determined by validation. Therefore, the validation process should carry out challenge test in order to make a proper evaluation for deviation which occurred in the operation process.

8.6.2 Revalidation of freeze drying technical process

When the following circumstances occurred, some contents of the validation should be revised according to the validation plan to confirm adaptation and reproducibility of the manufacturing process.

(1) Equipment and devices have great change or major maintaince.

(2) Changes of manufacturing techniques, testing methods.

(3) In the daily validation, such important technical parameters are unformal, public facilities or the quality of preparation are abnormal and so on.

8.6.3 Retrospective validation

Prospective validation is needed when the equipment is in initial operation, or in the design and beginning of production for a new products. Then the conventional production process begin. The technical process should have perfect daily monitoring plan so that the operation will be consistently in the situation required by validation.

Retrospective validation analyze a large number of recorded data and compared with the verification test results. It is emphasized that the original manufacturing record is not only an document of evidence but also the technical information.The retrospective validation finds a more suitable validation plan and the clue of improving process conditions by a comparative analysis of the recorded data in the original manufacturing process.

Freeze-drying technical process is complex and risky. The production of new products, the new technical process, the update of equipments or the change of technical parameters need prospective validation. Prospective validation cannot be replaced by retrospective validation.

References

[1] FLOSDORF E W, MUDD S. Procedure and apparatus for preservation in "lyophile" form of serum and other biological substances. J. Immunology., 1935, **29**: 389—425

[2] REY L, MAY J C. Freeze-drying/lyophilization of pharmaceutical and biological products. New York: Marcel Dekker Inc., 2004

[3] OETJEN G W, HASELEY P. Freeze-Drying. Weinheim :Wiley-VCH, 2004

[4] HUA T C. New technology of freeze-drying. Science Press, Beijing,2006

[5] ZHAO H G, LIN X C. Technology of freeze-drying. Huazhong Science & Technology University Press, Wuhan,China,1990

[6] HUA T C, LI Y F, LIU B L. The Principle and equipment of food's refrigeration and cold storage. China Machinery Press, Beijing,China,1999

[7] XU C H, ZHANG S W, GUAN K Z. Vacuum drying. Chemical Industry Press, Beijing,China,2004

[8] GAO F C. Freeze-dried food. China Light Industry Press, Beijing,1998

[9] LU B. New dosage form and technology of pharmaceuticals. People's Medical Publishing House, Beijing, 2000

[10] XIAO H H, LI J, HUA T C. Research on freeze-drying of the nucleated cell of human cord blood. Chinese Journal of Cell Biology, 2003, **25**(6): 389—393

[11] XIAO H H, HUA T C, LI, J, GU X L, Wang X. Freeze-drying of mononuclear cells and whole blood of human cord blood. CryoLetters, 2004, **25**(2): 111—120

[12] LI J, HUA T C, GU X L, XIAO H H. Morphological study of freeze-drying mononuclear cells of human cord blood. CryoLetters, 2005, **26**(3): 193—200

[13] HUA T C, REN H S. Cryobio-medical Techniques. Science Press, Beijing, China, 1994

[14] ZHOU G D. Structure and Physical Properties (the second edition). Higher Education Press, Beijing, 2000

[15] ASHRAE. ASHRAE Handbook (fundamentals). New York: 1993

[16] FENNEMA O R. Food Chemistry. 3rd edition. New York : Marcel Dekker Inc., 1996

[17] SLADE L, LEVINE H. Glass Transitions and Water Food Structure Interactions. In: Kinesella J E. Advances in Food and Nutrition Research. **38**. San Diego: Academic Press, 1994

[18] HUA T C. Cryobio-medical techniques: Study on Amorphous Solidification by fast cooling in cryobiomedical fields. in Thermophysical Studies in Biomedical Field edited by Chen H Q, Wang C C., Science Press, Beijing, 1995. 96—104

[19] LIU Z J, HUa T C, CHEN J M, GAO S. Effect of vitrification on freeze -drying process of pharmaceuticals. Chinese journal of pharmaceuticals, 2000, **31**(8): 380—383

[20] BLANSHARD J M V, LILLFORD P J. The Glassy State in Foods. England: Nottingham University Press, 1993

[21] ZUO J G, HUA T C, LIU B L, ZHOU G Y. Study of the freezing properties of solutions for freeze-drying by DSC. Chinese Journal of Cryogenics, 2005,(3):48—51

[22] HUA Z Z. Cryobio-medicine and thermophysics. Physics and Engineering, 2001, **11** (6): 1—4

[23] MERYMAN H T. Cryobiology. New York: Academic Press, 1966

[24] LUYET B. On the possible biological significance of some physical changes encountered in the cooling and the rewarming of aqueous solution. In: Asahina E. Cellular injury and resistance in freezing organisms. Japan: Hokkaido University Press, 1985. 1—20

[25] WU S Y, YAO K M, HUA T C. Study on cryobiological microscope system. Chinese Journal of Scientific Instrument, 1988, **9**(1): 90—93

[26] WU S Y, HUA T C. A new structure of cryomicroscope. Cryobiology, 1987, **6**: 553—554

[27] LIU B L, HUA T C , YUAN S M, WU S Y. A new type of cryo-microscope DSC system. Chinese Journal of Scientific Instrument, 1999, **20**(1): 60—62

[28] TAO L R, HUA T C. A microscopic study of the crystallization in cryoprotectant agents. Chinese Journal of Engineering Thermophysics, 2001, **22**(4): 481—484

[29] TAO L R, HUA T C. Microscopic Study of Crystal Growth in Cryopreservation Agent Solutions and Water. Annals New York Academy of Sciences, 2002, **972**, 151—157

[30] SUN Z Q, HUA T C. Research on bovine embryo's cryopreservation process by cryo-microscope .Chinese Journal of Refrigeration,1987,(10):25—27

[31] CHEN X H, HUA T C. Preliminary study on cryo-microscopic observation and preservation effect of the human leukocytes. Chinese Journal of Cryogenics, 1987, (4): 1—8

[32] WANG D R, HUA T C. Influence of unfrozen fraction to the survival of leucocytes during slow cooling. Cryobiology, 1988, **25**: 510

[33] HUA T C, WANG D R. The phase diagrams of ternary solution system in cryobiology. Chinese Journal of Refrigeration, 1989, (2): 11—20

[34] CHEN E T, ZHAO L, Hua Z ZH. A cooling and temperature controlling device for cryobiology. Chinese Patent : No.91-2-15370.9

[35] CHEN R T, HUA T C. A new cooler with very low consumption of liquid nitrogen. Cryobiology, 1988, **22**: 549—550

[36] CHEN R T, HUA T C. A new cryocooler for cryopreservation, Chinese Cryogenics, 1989, (2):24—29

[37] WU S Y, HUA T C. A new cooling technique in cryopreservation of biomaterial. Chinese Journal of Refrigeration, 1987, (4): 11—17

[38] PENG H Z, HUA T C, TAO L R. A programmable cooler with liquefied nitrogen vessel. Chinese Patent : ZL00-2-18101.0

[39] PENG H Z, HUA T C, TAO L R. A cryocooler with internal heating and pressing liquefied nitrogen. Chinese Cryogenics, 2000, (5): 8—12

[40] HUA T C, LIU B L. Heat and mass transfer in vitrification of bio- and food Material. Heat Transfer 1998, Proceeding of 11th Intern Heat Transfer Conf., Vol.1, August 23—28, 1998, Kyongju, Korea.

[41] REN H S, WEI Y, HUA T C. Theoretical predication of vitrification and devitrification tendencies for cryoprotective solutions. Cryobiology, 1994, **3**: 47—56

[42] HAN R H, HUA T C. Experimental investigation of cooling rates of small samples during quenching into subcooling LN_2. CryoLetters, 1995, **16**: 157—162

[43] CAO Q, HUA T C. Effects on rapid cooling of small samples in quenching. Annals of the New York, Academy of Sciences, 1998, **858**: 262—269

[44] XU J J, HUA T C. Comparative study on the quenching boiling of small and large samples in nitrogen. Chinese Journal of Cryogenics,1997, (4):15—18

[45] XU J J, HUA T C, LIU B L, FU X J. Experimental study on quenching boiling heat transfer from plate in subcooled liquid nitrogen. Chinese Journal of Cryogenics. 1999, (4): 174—179

[46] HUA T C, XU J J. Quenching boiling in subcooled liquid nitrogen for solidification of aqueous materials. Materials Science & Engineering A ,2000, **292**: 169—172

[47] LI Y F, HUA T C, LIU B L. Research on energy saving technologies in vacuum freeze drying of foods. Chinese Journal of Cryogenics,1997,(4):51—56

[48] SU S Q, DING Z H, HUA T C. Study on the measurement of the product temperature at the moving sublimation interface during the freeze drying process. Chinese Journal of Engineering Thermophysics,2004,**25**(5):852—854

[49] SU S Q, HUA T C. Method and device of the sublimation temperature during freeze drying Chinese Invention Patent: ZL 03 1 29506.1

[50] SU D S, WANG S L. Physical pharmaceutics. Chemical Industry Press, Beijing,2004

[51] GRAZIO F, FLYNN K. Lyophilization closures for protein based drugs. J. Parent. Sci. Technol., 1992, **46**: 54—61

[52] FRANK F. Long-term stabilization of biologicals. Biotechnology, 1994, **12**: 253

[53] HANCOCK B C, ZOGRAFI G. Characteristics and significance of the amorphous state in pharmaceutical systems. J. Pharm. Sci., 1997, **86**: 1—12

[54] SCHMITT E A, LAW D, ZHANG G. Nucleation and crystallization kinetics of hydrated amorphous lactose above glass transition temperature. J. Pharm. Sci., 1999, **88**: 291—296

[55] GOLDBLITH S A, REY L. Freeze drying and advanced food technology, London, Academic Press, 1975

[56] MUJUMDAR A S. Handbook of Industrial Drying. New York: Marcel Dekker Inc., 1987

[57] SANDALL O C, KING C J,WILKE C R. The relationship between transport properties and rates of freeze-drying of poultry meat. AIChE (American Institute of Chemical Engineers) Journal,1967,**13**(3): 428—438

[58] HELDMAN R, SINGH R P. Food process engineering. 2nd ed. USA: AVI Publishing Company, 1981

[59] TOLEDO R T. Fundamentals of food process engineering. 2nd ed. New York: Van Nostrand Reinhold, 1991

[60] SHEEHAN P, LIAPIS A I. Modeling of the primary and secondary drying stages of the freeze drying of pharmaceutical products in vials: numerical results obtained from the solution of a dynamic and spatially multi-dimensional lyophilization model for different operational policies, Biotechnology and Bioengineering, 1998, **60**(6), 712—728

[61] SADIKOGLU H, LIAPIS A I. Mathematical modeling of the prime and secondary drying stages of bulk solution freeze drying in trays: Parameter estimation and model

discrimination by comparison of theoretical results with experimental data, Drying Tech.,1997, **15**, 791—810

[62] BRUTTINI, R, ROVERO G and BALDI G. Experimentation and modeling of pharmaceutical lyophilization using a pilot plant, Chemical Engineering Journal, 1991, **45**: B67—B77,1991,

[63] MILLMAN M J, LIAPIS A I, and MARCHELLO J M, An analysis of the lyophilization process using a sorption-sublimation model and various operational policies, AICHE J.,1985, **31**: 1594—1604

[64] LI Y F, HUA T C. Characteristics of freeze-drying under the mass diffusion control. Chinese Journal of Refrigeration, 1997, (3):23—27

[65] SADIKOGLU H, LIAPIS A I. Optimal control of the primary and secondary drying stages of bulk solution freeze drying in trays, Drying Tech, 1998, **16**(3－5), 399—431

[66] MELLOR J D. Fundamentals of freeze drying. London: Academic press, 1978

[67] VAN ARSDEL W B, COPLEY M J, MORGAN A I. Food Dehydration.2nd ed. USA: AVI Publishing Company, 1973

[68] DA D A. Manual of vacuum design (the Third Edition). National Defense Industry Press, Beijing ,2004

[69] http://www.toyo-engi.co.jp/vf_c.htm

[70] http://www.tofflon.com/system.htm

[71] YU S M, BAI J, MA G Q. Preservation and freeze-drying technology of fruits and vegetables. Heilongjiang Science and Technology Press, Harbin, China, 1999

[72] FU W Y. Application of freeze-drying technology in instant coffee production, Chinese Journal of Packaging and Food Machinery, 2000,**28**(4):19—21

[73] ZENG Q X, XU X L. Hazard Analysis and Critical Control Points in Food Production. South China University of Technology Press, Guangzhou, China 2001

[74] YANG J B, WANG J, WANG B Q. Food Safety. China Light Industry Press, Beijing,1999

[75] Certification and Accreditation Administration of the People's Republic of China. Establishment and Implementation of HACCP system for fruit and vegetable juice. Intellectual Property Publishing House, Beijing, China, 2002

[76] WU Z Z. Application of HACCP system to the device and production of Freeze drying of kiwifruit. Bachelor's Thesis of Shanghai University of Science and Technology,2003

[77] CARPENTER J F, CROWE J H, CROWE L M. Stabilization of phosphofructokinase with sugars during freeze-drying: characterization of enhanced protection in the presence of divalent cations. Biochem. Biophys. Acta, 1987, **923**: 109—115

[78] ARAKAWA T, PRESTRELSKI S J, KINNEY W, CARPENTER J F. Factors affecting short-term and long-term stabilities of proteins. Advanced Drug Delivery Rev., 1993,**10**:1—28

[79] PRESTRELSKI S J, TEDESCHI N, ARAKAWA T, CARPENTER J F. Dehydration-induced conformational changes in proteins and their inhibition by stabilizers. Biophys. J., 1993, **65**: 661—671

[80] IZUTSU K, YOSHIOKA S, TAKEDA Y. The effect of additives on the stability of freeze-dried β-galactosidase stored at elevated temperatures. Int. J. Pharm., 1991, **71**: 137—146

[81] LESLIE S B, ISRAELI E, LIGHTHART B, et al. Trehalose and sucrose protect both membranes and proteins in intact bacteria during drying. Appl. Environ. Microbiol. 1995, **61**: 3592—3597

[82] LEVINE H, SLADE L. Another view of trehalose for drying and stabilizing biological materials. Biopharm., 1992, **5** (5): 36—40

[83] ALLISON S D. Effects of drying methods and additives on structure and function of actin: mechanisms of dehydration-induced damage and its inhibition. Arch. Biochem. Biophy., 1998, **358**: 171—181

[84] CARPENTER J F, CROWE J H. An infrared spectroscopic study of the interactions of carbohydrates with dried proteins. Biochemistry, 1989, **28**:3916—3922

[85] PIKAL K A, CARPENTER J F. pH changes during freezing in sodium and potassium phosphate buffer systems in the presence of glycine: effect on protein stability. Pharm. Sci.,1998,1(Suppl.):544

[86] CHANG B S, REEDER G, CARPENTER J F. Development of a stable freeze- dried formulation of recombinant human interleukin-1 receptor antagonist. Pharm. Res., 1996, **13**:243—249

[87] DELUCA P P, KAMAT M S, KOIDA Y. Acceleration of freeze-drying cycles of aqueous solutions of lactose and sucrose with tertiary butyl alcohol. Int. Technol. Pharm.,1989, **1**: 439—447

[88] WITTAYA-AREEKUL S, NAIL S L. Freeze-drying of tert-butyl alcohol/water cosolvent systems: A case report on formation of a friable freeze-dried powder of tobramycin sulfate. J. Pharm. Sci., 2002, **91**:1147—1155

[89] ZUO J G, HUA T C. Freeze drying of organic solution, Chinese Journal of Science and Technology of Food Industry,2006,(5):203—205

[90] KASRAIAN K, DELUCA P P. The effect of tertiary butyl alcohol on the resistance of the dry product layer during primary drying. Pharm.Res., 1995,**12**: 491—495

[91] PRESTRELSKI S J, PIKAL K A, ARAKAWA T. Optimization of lyophilization conditions for recombinant human interleukin-2 by dried-state conformational analysis using Fourier-transform infrared spectroscopy. Pharm.Res.,1995,**12**:1250—1259

[92] TOWNSEND M W, DELUCA P P. Stability of ribonuclease A in solution and the freeze-dried state. J.Pharm.Sci.,1990, **79**:1083—1086

[93] IZUTSU K, YOSHIOKA S, TERAO T. Decreased protein-stabilizing effects of cryoprotectants due to crystallization. Pharm.Res.,1993,**10**:1232—1237

[94] CHANG B S, RANDALL C S. Use of subambient thermal analysis to optimize protein lyophilization. Cryobiology, 1992, **29**:632—656

[95] SARCIAUX J M, HAGEMAN K J. Effects of bovine somatotropin (rbSt) concentration at different moisture levels on the physical stability of sucrose in freeze-dried rbSt/sucrose mixtures. J. Pharm. Sci., 1997, **86**:365—371

[96] KREILGAARD L, FROKJAER S, FLINK J M, RANDOLPH T W, CARPENTER J F. Effects of additives on the stability of humicola lanuginosa lipase during freeze-drying and storage in the dried solid. J. Pharm. Sci.,1999, **88**:281—290

[97] LIU Z H. Thermal Analyses—Book 8 of Analytical Chemistry Manual, Chemical Industry Press, Beijing, China, 2000

[98] Pyris Diamond DSC User Manual. 2002. Perkin Elmer Instruments, USA

[99] ZUO J G, HUA T C, LIU B L, ZHOU G Y. Thermal analysis of tertiary butyl alcohol/sucrose/ water ternary system. CryoLetters, 2005, **26** (5): 289—296

[100] ZUO J G, HUA T C. Annealing characteristics of lyoprotectant solution. Acta Physico-Chimica ,2005,**21**(10):1175—1179

[101] ZUO J G, HUA T C. Cryo-microscopic and thermal analyses of freeze drying solution,. Chinese Journal of Engineering Thermophysics,2006,**27**(2):307—309

[102] GAO C, ZHOU G. Y, XU Y, HUA T C. Freezing characteristics of EG and glycerol aqueous solutions studied by DSC. Acta Physico-Chimica Sinica, 2004. **20**(2):123—128

[103] Hutchinson J M. Studying the glass transition by DSC and TMDSC. J. Therm. Anal. Cal., 2003,**72**: 619—629

[104] GAO C, ZHOU G Y, XU Y , HUA T C. Glass transition and enthalpy relaxation behaviors of ethylene glycol and its aqueous solution with the presence of crystals. Thermo-Chimica Acta ,2005,**435**:37—42

[105] GAO C, WANG W H, HU T J, XU Y, ZHOU G Y, HUA T C. Glass Transition and enthalpy relaxation behavior of ethylene glycol and it's aqueous solution with different crystallinity. Acta Physico-Chimica Sinica,2004,**20**(7):701—706

[106] GAO C, ZHOU G Y. Phenomenological study of structure relaxation of glycerol aqueous glass. Acta Physico-Chimica Sinica,2005,**21**(8):909—914

[107] GAO C, ZHOU G Y, XU Y, HU T J, HUA T C. Glass transition and structure relaxation parameters of glycerol/water mixtures with high concentrations: A DSC study. Chinese Journal of Chemical Physics,2005,**18**(3):457—462

[108] ZUO J G. HUA T C, LIU B L, ZHOU G Y, XU Y. Assay study on glass transition temperature of solutions in freeze-drying. Chinese Journal of Food Science, 2006, **27**(2): 58—60

[109] HATAKEYAMA T, LIU Z H. Handbook of thermal Analysis.1st ed. England: Wiley,1998

[110] GUO Y. Bio-pharmaceutical Technology. China Light Industry Press, Beijing,2000

[111] WANG P S, WANG Y M. Status and development of biopharmaceutical industry in the world. Chinese Journal of Fine and Specialty Chemicals,2003,(5):8—9;(6): 8—11

[112] LIU Z J, HUA T C. The denaturation mechanism of protein pharmaceuticals in freeze-drying process. Chinese Journal of Biochemical Pharmaceutics, 2000, **21**(5):263—265

[113] O'BRIEN F J, HARLEY B A, YANNAS I V. Influence of freezing rate on pore structure in freeze-dried collagen-GAG scaffolds. Biomaterials. 2004, **25**(6): 1077—1086

[114] STRAMBINI G. Phosphorescence lifetime measurements indicate ice-induced partial unfolding of proteins. J. Biophys, 1996, **70**: 971—976

[115] TOWNSEND M W. Use of lyoprotectants in the freeze-drying of a modal protein ribonuclease. PDA J. Pharm. Sci. Tech., 1998, **42**(6): 190—196

[116] CHEN L Z, LI Y G, BIAN G Z. Protective effect of trehalose on freeze-dried products containing enzymes. Shanghai Journal of Medical Laboratory Sciences, 2001, **16**(3):166—167

[117] GE Y, LIU L, GAO J J. Protective effect of trehalose on recombinant human copper and zinc superoxide dismutase (rhCuZn-SOD). Chinese Journal of Biological Products,2003,**16**(1):42—45

[118] NAIL S, LIU W, WANG D Q. Loss of protein activity during lyophilization as an interfacial phenomenon: formulation and processing effects. Freeze-Drying of Pharmaceuticals and Biologicals, August 1 - 4, 2001, The Village at Breckenridge, Colorado, USA

[119] HORA M S, RANA R K, WILCOX C L. Development of lyophilized formulation of interleukin. Developments in biological Standardization, 1992, **74**: 295—306

[120] ARAKAWA T, PRESTRELSKI S J, KENNEY W C. Factors affecting short-term and long-term stabilities of proteins. Adv. Drug Deliv. Rev., 2001, **46**: 307—326

[121] CARPENTER J F, PRESTRELSKI S J, DONG A. Application of infrared spectroscopy to development of stable lyophilized protein formulations. Eur. J. Pharm. Biopharm. 1998, **45**(3):231—239

[122] QI Y Y. Protective agent's screening and freeze-drying technology of human leukocyte interferon Acta Universitatis Medicinalis Nanjing,1997,**17**(3):284—286

[123] BYEONG S C, ROBERT M B, AICHUN D. Physical factors affecting the storage stability of freeze- dried Interleukin-1 receptor antagonist: Glass transition and protein conformation. Archives Biochem. Biophys., 1996, **331**(2): 249—258

[124] SU S Q. Parameter measurement in the freeze-drying process and freeze-drying experiment of biological pharmaceuticals. Doctoral dissertation of Shanghai University of Science and Technology,2004

[125] PING Q N. Modern Pharmaceutics. China Medical Science and Technology Press. Beijing,1999

[126] BI D Z. Pharmaceutics. People's Medical Publishing House, Beijing,1999

[127] ZHU S S. New dosage form of pharmaceuticals. People's Medical Publishing House, Beijing,1993

[128] GOMEZ-HENS A, FERNANDEZ-ROMERO J M. Analytical methods for the control of liposomal delivery systems. Trends Anal Chem., 2006, **25**: 167—178

[129] MOZAFARI M R & MORTAZAVI S M. Nanoliposomes: From Fundamentals to Recent Developments. Trafford Publishing Ltd, Oxford, UK, 2005.

[130] HUA T C, LI B G, LIU Z Z , SUN D W. Freeze-drying of liposomes with cryopetectants and its effect on retention rate of encapsulated florafur and vitamin A. Drying Technology, 2003, **21** (8):1491—1505

[131] SU S Q, HUA T C, DING Z H, HU S G, HAN L. Effects of lyoprotectants and rehydrated solutions on the encapsulation of HB- Ⅰ a liposomes. Chinese Journal of New Drugs, 2004, **13**(9): 809—812

[132] SU S Q, HUA T C, DING Z H, HU S G, HAN L. Vesicles size and diameter distribution of HB- Ⅰ a liposomes during freeze-drying. Chinese Journal of Pharmaceuticals,2004,**35**(3):154—157

[133] LIU Z J, HUA T C, TAO L R, LIU B L. The effect of cryoprotectant concentration and cooling rate on the glass transition temperature of liposome. Chinese Journal of Cryogenics,2000,**118**(6):1—5

[134] LIU Z J. Experimental study and mechanism analysis on lipofectin drug during the low temperature freeze-drying. Doctoral dissertation of Shanghai University of Science and Technology,2001

[135] LIU Z J, HUA T C, LI B G. Experimental study on liposome Freeze-drying Process. Chemical Industry and Engineering Progress, 2000, **19**(11): 69—72

[136] LIU Z J, HUA Z Z. The effect of the optimal freeze-drying cycle on the quality of liposome . Millennium international Symposium on Thermal and Fluid Sciences, Xi'an, China, September 18—22, 2000

[137] LIU Z J, XIAO H H. SU S Q, HUA T C. The experimental study on the influence of freezing method on the size and encapsulation efficiency of the freeze-dried liposome pharmaceuticals. Chinese Journal of Engineering Thermophysics, 2002, **23**(5): 599—601

[138] LU D Y. Medical Microbiology (the fourth edition). People's Medical Publishing House , Beijing,1995

[139] JI X H, ZHANG J Q. Medical Immunology and Medical Microbiology.Science Press,Beijing,2001

[140] YE Y Y. Guidelines of pharmaceutical production. Chinese Medical Science and Technology Press, Beijing,1996

[141] FREDERICK J C, Agalloco J P. Validation of Aseptic Pharmaceutical processes. New York: Marcel Dekker, Inc., 1986

[142] China Pharmacopoeia Committee. Pharmacopoeia of the Peoples Republic of China, (2005 English Edition), People's Medical Publishing House, Beijing, 2005

[143] Chinese Biological Products Standardization Committee. Chinese Regulations of the Biological Products. Chemical Industry Press, Beijing, 2000

[144] BAI H L, LI W C. Verified guidelines of pharmaceutical products. Chemical Industry Press, Beijing,2003

[145] China Pharmaceutical Industry Association, China National Pharmaceutical Industry Corporation. The Guideline of Good Manufacture Practice for pharmaceutical products. Chemical Industry Press, Beijing, 2001

[146] FREDERICK J, CARLETON J P. Agalloco, Validation of Pharmaceutical processes—sterile products. New York :Marcel Dekker, Inc., 1999

[147] http://en.wikipedia.org

Index

S

T

FSC
www.fsc.org
MIX
Papier | Fördert gute Waldnutzung
FSC® C083411

Kolibri 360 GmbH
Ferdinand-Jühlke-Straße 7
99095 Erfurt, Deutschland
produktsicherheit@kolibri360.de